SERVICE SECTOR WORKERS IN A MANUFACTURING CITY

Service Sector Workers in a Manufacturing City

A Study of Work Histories and Attitudes in a Coventry Service Industry

IAN PROCTER

Department of Sociology
The University of Warwick

Avebury

Aldershot · Brookfield USA · Hong Kong · Singapore · Sydney

Published by

Avebury

Gower Publishing Company Limited,
Gower House, Croft Road, Aldershot,
Hants. GU11 3HR, England

Gower Publishing Company,
Old Post Road, Brookfield, Vermont 05036
USA

British Library Cataloguing in Publication Data

Procter, Ian, *1947*-
Service sector workers in a manufactoring city : a study of work histories and attitudes in a Coventry service industry.
1. Service industries workers —— Great Britain
I. Title
331.12'51'0941 HD8039.S452G7

ISBN 0 566 05618 6.

Printed and bound in Great Britain by
Athanaeum Press Limited, Newcastle-upon-Tyne

Contents

List of tables

Acknowledgements

I owe a debt of gratitude to the management and trade union representatives of the Coventry (Lanchester) Polytechnic for giving permission for me to undertake this study. In particular, Rita Adams gave me great assistance in providing a sampling frame of employees, introductions to supervisors and arranging interviewing rooms in the Library and Careers Service. Wherever I went in the Polytechnic I was impressed by the patience and helpfulness of the staff which facilitated our field work enormously. To the people who participated in the study I obviously owe a special debt as it is the information they gave us which is reported here.

At the University of Warwick, I would like to thank the Research and Innovations Fund for providing the grant which enabled this project to go ahead. I was assisted by comments on papers I presented at the Department of Sociology Research Day, May 1986, and the Faculty Labour Studies Group. My colleagues Peter Elias and Peter Fairbrother gave valuable help in designing the questionnaire and clarifying my ideas at an important stage of the analysis. I owe particular thanks to Mo Padfield who acted as my research assistant at the time of the field work and data entering. Mo's enthusiasm, conscientious work and perceptive observations were a boon throughout.

Finally, in more personal terms, Ed, James, Joe and Bess keep me sane by driving me mad, and sincere thanks to Jill Allen for constant encouragement and support in every way.

1 Introduction

The context of the study.

The following report is based upon interviews with sixty-seven people who were employed as ancillary workers and technicians by the Coventry (Lanchester) Polytechnic in the autumn of 1985. Their particular significance from the point of view of the research project was that they worked in an industry in what is generally called the 'service' sector of the economy, an organisation geared to providing a service rather than a material product as in the extractive and manufacturing sectors.[1] The last decade or so has seen a number of important changes in the structure of the British economy, the return to mass unemployment being the most prominent. Yet it is also recognised that this in itself is intimately connected to a second change, the decline of employment in the secondary or manufacturing sector and the increasing prominence of service industries.[2] Data from the 1971 and 1981 Census of Great Britain can illustrate this.[3] In 1971 8,430,740 people were employed in the manufacturing and construction industries, by 1981 this had fallen to 6,858,690, a decline of over a million and a half people, 18.6% of those employed in this sector in 1971. On the other hand in 1971 the service sector employed 12,539,800 but by 1981 this had increased to 13,892,800, a further one and a third million people, or 10.8% of the 1971 figure.[4]

This national trend toward the decline of secondary industrial employment and increasing employment in services is found in the local labour market which provided the setting for this study, the City of Coventry and its immediate surroundings. In 1971 111,010 people worked in manufacturing and construction, by the 1981 census this had fallen by fully 36,080 to 74,930, a decline of 32.5%. Service sector employment increased by 6,740, from 61,620 to 68,360, a rise of 10.9%.[5] Two points are of note concerning Coventry in relation to the national figures just quoted. The decline in secondary sector employment in Coventry has been far steeper than in the country as a whole. The national figure is 18.6%, Coventry's 32.5%.[6] Secondly, although the Coventry service sector increased by the same proportion as the national picture over the census decade, it is clear that, compared to the national economy, the Coventry service sector was and remains relatively small. In 1971, in Great Britain, 52.8% of the employed population worked in service industries but in Coventry service employment accounted for only 34.8% of total employment. In 1981 the services employed 60.6% of Great Britain's employed population but the figure for Coventry was 53.3%. What these bare statistics reflect is the particular character of the Coventry local economy, an economy historically dominated by manufacturing industry.[7] Precisely because of this, the national trend of decline in manufacturing over the last twenty years or so has been especially marked in the City, for throughout this century the City's growth has been on the basis of its staple industries of vehicles, electrical and mechanical engineering, aerospace and artificial fibres, that is, just the industries which have shed labour so drastically in the recent past. Despite this contraction Coventry remains a manufacturing city[8] but the fact remains that the balance of the local economy has shifted dramatically towards the service industries in the last decade and a half. To repeat, in the early seventies only around a third of the local working population were to be found in services, now the figure is over half and it exceeds the number employed in the secondary sector which, until the early seventies, employed around seven out of ten Coventry workers.

We can thus see that the national trend has taken a particularly pronounced form in Coventry because of its manufacturing base and the severity of employment decline in that sector. But so far this discussion has been cast entirely in what might be called *aggregate* terms, that is, taking the total numbers employed in the different sectors as the measure of economic change. This certainly gives us the context of change but it does not tell us about how individuals who have lived *through* this change have experienced it. What has happened to the 36,000 people who worked in manufacturing in 1971 but who did not in 1981? Of course some of these people will have died, some will be retired, some will have moved to another area and far too many have become unemployed. The other possibility is that some individuals have moved from jobs in manufacturing to jobs in the service industries. Aggregate figures do not tell us whether this has occurred or not. We know that the total employed in manufacturing has declined and that the total employed in services has increased. But this does not mean that workers displaced from one sector are redeployed in the other. Indeed there are good reasons why this might not be the case. Many of those shed by manufacturing plants were male, full-time workers who were formerly employed in 'traditionally' male jobs such as track assembly workers or machinists in car factories. On the other hand, the service industries tend to employ relatively large numbers of women workers, often in a part-time capacity, and doing 'traditionally' women's jobs such as catering work.[9] Turning again to the Coventry census figures there seems some support for this view. We have noted that the secondary sector employed thirty-six thousand fewer people in 1981 as compared to ten years earlier. If we break this down by sex we find that these industries employed 28,200 fewer men and 7,880 fewer women, that is, the great majority of people displaced from manufacturing were indeed men. In the service sector, on the other hand, only a meagre 580 of the total increase in the workforce were men, overshadowed by an increase of 6,160 women. Yet once again these aggregate figures cannot in themselves answer our question as to where individuals have gone over the period of economic change that we are considering? For example, more women were displaced from manufacturing than their increase in services so that it might be the case that there has been a shift of women workers from one sector to another. Again, although the increase in male workers in services has been very small, we have no idea whether those employed in the services in

1981 are the same people who worked there in 1971. It might be that male employment in services is slanted to jobs for older men, and, as these come up to retirement, they are replaced by men moving from manufacturing jobs. The simple point is that aggregate data cannot provide answers to questions which refer to the movements of individuals over time. For this, another research technique is required which maps out the employment biography of individuals.

The research site.

This was the starting point of this study, to take a group of workers who are *currently* employed in service industry and to investigate what they had done throughout their working lives to discover whether or not they had moved from the secondary sector. The service sector is notorious for the variety of industries which are grouped together under its banner and the range of different working environments it encompasses. It includes, on the one hand, the small corner shop and the one person business such as a self-employed plumber or hairdresser, on the other, the massive bureaucracies of the public services and private institutions such as banks and building societies. Given the limited resources of this study, it was clear that the full range of service industries could not be covered, indeed the focus would have to be on one particular context. The workplace selected needed to meet two criteria. First, it should have been expanding in employment terms over the last fifteen years or so, the period of contraction in manufacturing work. The expansion of services has not been at all even, some have remained static or even declined, notably distribution and transport.[10] Second, the kind of work available should have a *degree* of compatibility with work in manufacturing industry, so that it would be at least feasible for workers displaced from the latter to find employment in this new environment. This tends to exclude the commercial division of banking, accountancy and the like which offer few opportunities to manual workers leaving manufacturing. This narrows the field to the public services which have grown considerably over the recent past[11] and provided a range of manual jobs such as cleaning, catering, portering, maintenance work and technical posts in laboratories. The Polytechnic seemed a suitable site for the investigation because, unlike a particular school, for example, it was a large workplace and in this respect similar to the large factory which was characteristic of Coventry manufacturing. As against other large workplaces, such as a hospital, its science and engineering laboratories might offer employment opportunities for skilled labour leaving local industry.

The Polytechnic thus seemed an appropriate workplace in which to examine the first problem of the research, the degree of transfer from declining secondary employment to expanding service sector employment. It was also a suitable site to explore a second area, already hinted at in the above paragraph, whether workers perceived a qualitative difference between employment in the two contexts of secondary industry and services? Some service sector workplaces are so different that this might be taken as almost self-evident. But what of the large establishment, with a distinctly manual workforce and extensive unionisation of the workforce? Was this a markedly different environment to the manufacturing plant? The Polytechnic is a large organisation, as well as the 621 people in the occupations of interest here,[12] it has an academic staff of some 500, a considerable administrative and clerical workforce and, of course, a student body of some five and a half thousand.[13] The workers included in this project are in some cases doing distinctively manual jobs such as cleaning and catering. In other cases this is not so self-evident, the protection officers are monthly paid 'staff' and their work is to provide a service (providing for the security of the buildings and their occupants). Yet, as we shall see, men doing this work required no formal qualifications and often came from manual employment in manufacturing. The technical staff are in some ways not 'manual' workers, yet their very presence in academic departments in which the teaching staff have very different conditions of employment, gives them a sense of separateness from the people who are in a sense their immediate managers. Finally, the Polytechnic subscribes to the following statement of City Council policy on union membership, and as we shall see most employees are union members.

> 'The City Council, believing in the principle of collective bargaining and being in membership of the joint organisation of employees and employed, encourages all their their employees to be members of an appropriate organisation.'[14]

For these reasons the Polytechnic is a suitable organisation to investigate the two research problems of the study, the extent to which it had drawn its workforce from the declining secondary sector and the workers' perception of similarity and difference between work in the secondary and service sectors. But it is also clear that the Polytechnic is a specific employment context and cannot be taken as typical or representative of the service sector as a whole. Indeed without comparative data its representativeness of large, unionised organisations in the public sector can only be surmised.

Summary.

A number of methodological issues are taken up in the substantive discussion but chapter two gives an outline of the approach adopted and a discussion of the questionnaire employed, so that the reader has an indication of the range of data available.

The research was concerned with the Polytechnic's ancillary and technical staff. In chapter three this workforce is described very briefly as a whole to show why the sample of interviewees was selected. A more detailed account is given of the sample of sixty-seven people to indicate their *current* characteristics. The main points here are that the workforce is relatively old, tends to be currently married and living in households with children. The spouses of the workers tend to be economically active. The sample first came to the Polytechnic in their thirties, typically via 'informal' means of obtaining a job. Once at the Polytechnic, people tend to stay for some time, although there is considerable movement from job to job.

These current characteristics of the workforce are only a background to the study because the intention was not to undertake a detailed investigation of the social relations of work but to look at the peoples' working lives and then their comparisons of different jobs held. Thus the next two chapters give an analysis of the work histories of the sixty-seven people, beginning, in chapter four, by looking at their history of work in overall or aggregate terms to see where they had spent their working lives. Clearly there are always individual idiosyncrasies in such biographies but there are also general paths which different groups of people follow, the most obvious being the distinction between men and women, men usually having a continuous participation in the labour market, many women withdrawing temporarily to bear and care for children. (Barrere-Maurisson, M.A. *et al* 1985, Martin, A. and Roberts, C. 1984) Chapter five offers an analysis of such different patterns of labour market experience over time, based upon the actual histories of the people in the sample and the main interests of this study. It was found that the work biographies of younger workers were distinct from older people in the sample so these form one group. Then the older workers are distinguished by sex for the reason mentioned above. Most of the older women had broken their labour market involvement for a period or periods of unpaid domestic work. Two had not done this so they are separated out to see whether any differences between them and the main group emerge. The older men are split into four groups. Most of them had indeed transferred to service industry after spending most of their previous working life in manufacturing or construction. However, the manner of transfer differed, for some it represented what we shall call a 'career development', for others a traumatic 'adjustment to employment crisis'. A minority of older men had not transferred from one sector to another, a few had moved back and forth throughout the working lives and a few had never worked in the secondary sector. Thus seven different patterns of labour market experience are described; two for older women distinguishing between those who had taken a 'domestic break' and those who had not, four for older men distinguished by their employment in the secondary sector and their move to service industry, and the recent experience of younger workers.

These different work history patterns are used in chapters six to ten as a framework to analyse the sample's estimations of 'best' and 'worst' jobs and their comparisons of these jobs with their present one. As mentioned above, the second objective of the research was to explore people's views about the different jobs they had held. The technique employed was to ask them to nominate their 'best' and 'worst' jobs. Clearly jobs vary in different respects, one might pay good wages but offer little opportunity for the person to work autonomously. Thus this best/worst technique was applied to a number of dimensions. Those selected were money, skill, job autonomy, trade union representation and best/worst jobs in overall terms. Each individual thus nominated ten jobs, the five best and five worst. Insofar as the selected job was not the present one, a further question was asked to establish a comparison between the selected job and the current one. The aims of these chapters were to establish which jobs were considered best and worst, whether these judgement varied with different labour market experience, and how jobs at the Polytechnic stood by comparison with others, particularly those in manufacturing and construction?

The report concludes with a detailed summary of the argument and a drawing together of the patterns of work history and respondents' estimations of their best and worst jobs.

Notes.

1. There is some debate as to quite how to differentiate the tasks and organisations that make up an economy into such sectors, see for example Jones, B. (1982) in which a number of different schemes are outlined and a fivefold sectoral division proposed. For the purposes of this report the threefold scheme of the primary sector (agricultural and extractive industries), secondary sector (manufacturing and construction) and tertiary sector (service industries) will be employed as this is the context in which the debate about the significance of the increasing prominence of the latter has been conducted.
2. See, for example, Handy, C. (1984), Gershuny, J.I. and Miles, I.D. (1985), Robertson, J.A.S. (1982).
3. The Census is used here to facilitate national-local comparisons and to enable further breakdown of data a little later. The main disadvantage in using it is that it refers to the decade coming to a close in 1981, that is in the middle of the recession which led to the displacement of labour on such a large scale. Thus the figures used here should be seen

as indicative of the trend rather than as precise measures of a process with a specific start and end point. Given the continuing impact of redundancy in the city it is highly likely that the figures used here underestimate the extent of job loss in Coventry's manufacturing industry. Healey and Clark put it at 46% between 1974 and 1982 (see footnote 6), the City Treasurer's Economic Unit (1986) at 51.8% between 1974 and 1984.

4. The sources used here are 1971 Census of Great Britain, Report on Economic Activity, Part II, Table 18, Pp. 196 and 1981 Census of Great Britain, Report on Economic Activity, Table 9, Pp. 204. The census employs the Office of Population, Census and Statistics **Standard Industrial Classification** which groups industries into ten divisions. Three divisions are included here as the secondary sector: 3.Metal goods, engineering, and vehicles; 4.Other manufacturing industries; 5.Construction. The service, or tertiary, sector is made up of four divisions: 6.Distribution, hotels and catering, repairs; 7.Transport and communication; 8.Banking, finance, insurance, business services and leasing; 9.Other services. Division 9 is mainly composed of the public services such as those provided by central and local government and the health service.

5. The sources here are 1971 Census of Great Britain, Economic Activity Leaflet for the County of Warwickshire, Table 3, and 1981 Census of Great Britain, Economic Activity, West Midlands, CEN 81 EA, Table 3. Note that the 1971 figures refer to the former County Borough which became the Metropolitan District in the 1974 reorganisation of local government. The District was enlarged by boundary changes but in population terms this was minimal and certainly did not influence the development under consideration here.

6. The rapid decline of manufacturing is documented by the research of Healey and Clark using The Coventry Region Industrial Establishment Databank, see Healey, M. and Clark, D. (1984). This research used a rather different time span than the census, (1974 to 1982), a geographical area rather larger than the local authority boundary, and did not include construction. The main finding of the study was that 'In 1974 115,000 people were employed in manufacturing; by 1982 this had fallen by 46% to 62,400.' (Healey, M. and Clark, D. 1983 Pp. 51)

7. The particularities of the Coventry economy are described and analysed in Richardson, K. (1972), Chapters 2 to 5; Rosser, M. and Mallier, T. (1981) and Thoms D.W. and Donelly, T. (1986).

8. Thus of 30 urban areas with a population of over 150,000 (excluding the London conurbation) Coventry has the third highest proportion of its employed population in manufacturing industry, next to Sandwell and Stoke-on-Trent. (Procter, I. 1984b).

9. See the study of the changing composition of the workforce by Massey, D. (1984).

10. In Coventry M.D. employment in the distributive sector increased by only 190, from 22,640 to 22,830 between 1971 and 1981 whilst employment in transport declined from 5,720 to 5,560.

11. In Coventry M.D. the 'Other Services' division of the service sector grew in employment from 28,460 in 1971 to 32,700 in 1981.

12. I.e. in ancillary and technicians posts, the workforce and the sample drawn from it will be discussed in chapter 3.

13. The figures for academic staff and students are taken from the Polytechnic's 1984 Development Plan and are for the academic year 1983/84 (Pp. 9 and 44). I have been unable to obtain exact figures for the non-academic support staff.

14. This is a standard statement taken from the further particulars of a job advertised at the Polytechnic in September 1986.

2 Methodology, the structure of the questionnaire

The scope of this chapter.

The reasons why the Polytechnic was selected as the research site have been explained in the Introduction. I approached the Polytechnic's management for permission to invite workers for interview. This was granted subject to agreement with the trade unions, which was forthcoming on the understanding that individual confidentiality would be respected, interviews would be conducted in 'works time' and that all the Polytechnic would supply to me was a list of names for sampling purposes. All of this was entirely satisfactory to me and the management had already agreed that people could be interviewed during their normal hours of work.

The main research instrument to be employed in the research was a structured questionnaire, used to interview a sample of the ancillary and technical workforce. It is appropriate to leave matters concerning the sample design until chapter three as this influences the structure of the sample which is discussed there. Similarly, a number of issues relating to respondents' understanding and response to questions will be discussed in the chapters on work history and 'best'/'worst' job estimations as these affect the quality of data available. However, at this point a general overview of the questionnaire employed is appropriate to indicate the kind of data that was obtained.

The questionnaire was designed during July and August 1985 and piloted during this period as various drafts were developed. Systematic piloting was difficult as it would have been virtually impossible to gain access to another establishment employing a similar kind of workforce. So most of the pilot interviews were conducted with friends of the researchers, (four in all). However two interviews were conducted amongst staff at my own place of work in their own time, one with a technician and the other with a cleaner. The main interviewing took place between September and December 1985. It was undertaken by the researcher and his assistant. Both interviewers had been involved with the design of the questionnaire and had done the pilots, so were familiar with its mechanics, logic and objectives. Interviewing took place during the respondent's normal hours of work, usually in rooms kindly provided by the Polytechnic's Careers Service and Librarian, although some were done where the individual actually worked if this was more convenient. In three cases the interview took place in the individual's own time. One person came in on a rest day, though this only became known to us during the course of the interview and the respondent was perfectly happy about it. One interview proved difficult to conduct at the actual place of work and the respondent invited the interviewer to complete the interview at home. This person then proved to be instrumental in arranging a further interview which also took place at this second respondent's house.

The reason why the above interview was difficult to conduct during works time was that the individual concerned worked with another person who naturally objected to doing all the work. This raises the more general point of the pressure of time in the interview situation. The Polytechnic's management had given permission for the interviews to be done during working hours and circulated a memorandum to supervisory staff asking for their cooperation. Such cooperation was willingly and patiently offered but it was noticeable that there were differences in the practical arrangements for covering the person being interviewed. The reason for this was that some workers had to do certain tasks at a certain time and replacements were simply not available, others could either rearrange their work or be covered by available colleagues. This difference largely ran along occupational lines although there were significant exceptions to each. On the one hand there were caterers and cleaners who either worked in small teams or had specified tasks to do and their interview time meant that either their colleagues were doing their work or the job was not being done. The upshot of this was that amongst these workers there was a sense of urgency in the interview, an awareness on the part of the interviewee that the work was still to be done or that workmates were having to do more. On the other hand were technicians, protection officers and attendants. In these cases the nature of their work meant that it could be delayed or someone could cover for them with the result that the interview was much more relaxed and the individual had much more time to offer information and opinion. As mentioned this difference tended to be linked to the occupation of the person but there were many exceptions. Its influence on the data collected was not significant because the responses to the formal questions did not differ, the variation occurred between the individual's propensity to expand on the actual information being asked, for example, to tell the interviewer not only when children were born and left school (see below) but also what they had subsequently gone on to do!

To move on to the overall structure of the questionnaire. As explained above, its main tasks were to ascertain the individual's work history and responses to questions on 'best' and 'worst' jobs, these are discussed in sections 4 and 5 below. The questionnaire also included a section on family history as it has become clear that it is impossible to interpret people's work biography, particularly but not exclusively womens', without knowledge of their family formation. (Barrere-Maurisson, M.A. *et al* 1985, Martin, A. and Roberts, C. 1984). Section 3 is concerned with this. Finally, the work history required a starting point, quite when an individual was deemed to begin 'work'. This is dealt with in the following section.

The starting point.

Before broaching this issue it should be noted that prior to the interview a number of basic items of information were already known about each respondent. The person's name, sex, full/part-time status and broad occupational category[1] were contained in the list of workers supplied by the Polytechnic. These items of information were included in boxes at the beginning of the questionnaire, completed before the interview began, along with the interview number, the date and time of the interview and the interviewer's name. A couple of points should be noted regarding this 'pre-interview' information.

The person's name was recorded on the questionnaire so that the particular individual could be identified and politely addressed at the time of the interview. However, when the interviewer returned the completed schedule to the office the corner of the page containing the individuals name was cut off and from then on the person became anonymous as their interview number. At the beginning of the interview the assurance of confidentiality given in the letter of invitation[2] was reiterated.

Although preliminary information was available regarding full/part-time status and the person's occupation this would be supplemented by further questions in the work history section. There a more detailed description of the person's job and status would be ascertained. In fact, in a number of cases, individuals designated as part-time on the list of names turned out to be full-time by the time the interview took place.

To come on to the question of when an individual's biography of work was deemed to commence. Clearly many young people become involved in paid employment prior to leaving the education system, working as shop assistants, delivering newspapers or helping about the house in exchange for their pocket money. (Finn, D. 1984). Again, perhaps most college students do some kind of vacation jobs, or at least make themselves available for paid employment during their time in higher education. However it was decided to ignore this labour market experience on practical, theoretical and definitional grounds. In the first place, ascertaining such information would be time consuming, particularly as it was strongly suspected that many of the respondents would be in middle age and over and their memory of school time jobs would be hazy. Second, it was not at all clear that such early employment experience was significant to an individual's work experience and this project was not the place to establish this one way or another. Finally, the parameters of labour market participation amongst the young are extremely blurred. Is helping with housework and receiving pocket money 'employment'? Is a student registering for work in the summer entering the labour market? Thus it was decided to make the starting point of the work history the time when the individual left full-time, continuous education.

On the whole this worked out well in practice but there were a small number of respondents in whose cases this criterion was changed. These were people who, whilst being students in higher education, had taken sandwich courses which had involved quite long (i.e. six months or a year) periods in paid employment. From the beginning of these interviews it was clear that the individuals concerned regarded the employment experience gained at the time as significant to them and it would have been perverse to ignore it.

The first set of questions established the person's age at the time of leaving full-time, continuous education, the type of school attended, qualifications obtained at this time, whether they were resident in the local area and their parents' occupation at this time. The work history thus had a starting point in terms of the individual's education, geographical location and family background insofar as this was indicated by parents' occupation and location in the class system.

Family history.

The main objective of the research was to establish people's work history. However, as mentioned already, on the basis of existing research, the view was taken that an individual's work history cannot be interpreted unless certain basic facts about their family responsibilities are taken into account. For women this is very well established, care of the home and family have to be balanced against the opportunities and constraints of employment. More tentatively, it has been suggested that men's employment becomes more demanding (e.g. in terms of overtime and shiftwork) when they are responsible for children. So it was essential to collect certain items of information about the individual's family responsibilities over the course of their working life. It was decided to include these questions prior to the work history because most people have a reasonable recollection of the dates of key events in their family history (e.g. of marriage and birth of children) which, once established, could be used as an *aide memoire* in establishing the chronology of work. This proved to be the case in practice, all our respondents could remember the year of the family events we were interested in. Most of the women could remember the month of the event, although many of the men could not, often commenting that their wife was responsible for remembering and acting on birthdays and anniversaries. The information on family history was recorded on a separate sheet which could be consulted during the build up of the work history and many respondents did utilise it.

The information required about family history was when (if ever) the individual was married and when they were responsible for children and any other dependents. 'Marriage' was not limited to legalised relationships. 'Responsibility' was assumed to begin either at the time of birth of a natural child or when the person adopted or became the step-parent to other children. In the case of other dependents, the respondent was asked when responsibility commenced and ended. The ending of responsibility for children is much more diffuse. It may be when a child leaves education, when the child reaches a certain age, or when the child leaves the respondent's home, either to live in their own household of some kind, or if the family breaks up through separation or divorce. Information was obtained to cover each of these possibilities for each child.

Work history.

This was the core of the questionnaire and a large amount of information was desired for two reasons. First, the term 'work' was intended in a wide sense and not restricted to paid employment. Again much recent research and commentary in this area has stressed the necessity to include as full a range of productive activity as possible. (For example: Allen, S. 1982; Brown R.K. 1984; Handy, C. 1984; Pahl, R.E. 1984). Whilst the amount of data that could be collected on non-paid work was limited and not directly relevant to the aims of the project, the existence of such activity and its place in an individual's overall work biography needed to be established. The generic term used in the project for an activity is 'work period'. This includes the following kinds of 'work':[3]

Employment.
Self-employment.
Unemployment.
Full-time education/training.
Domestic work.
Military service.
Long term sickness.
Voluntary work.
Other.

Second, for each period of employment it was intended to collect a comprehensive range of information about features of jobs which could be related to the individual's evaluations of different jobs. In view of this, it was decided to approach the work history in two stages. The first stage asked for certain basic items of information which could establish the chronology of work periods experienced by the individual since leaving education. This was entered on a separate sheet, the Work History Chronology, which could be consulted by both interviewer and interviewee in the second stage. During this, more detailed information about each work period was requested. This strategy of establishing the outline list of work activities and then filling in detail later worked well in most cases. The two exceptions will be noted later.

Stage one.

In the first stage of the work history the following information was obtained for each work period from leaving school or college to the present job.

The month and year when the work period began and ended,

The individual's economic status in terms of the categories above.

Insofar as the work period was paid employment:

Whether it was self-employment or employment.
The individual's occupation.
The name of the employer and the type of industry.
Whether the job was full or part-time and the weekly hours worked.
When the individual changed employer or the nature of the work changed.

A number of points are of note regarding the construction of the Work History Chronology.

a) On the interview schedule the questions were worded precisely and listed in a definite order to establish the information required and to move from one work period to another between school/college and the present job. However, in practice most interviews were not conducted in this tightly structured way. This was for a number of reasons. First, the Work History Chronology was laid out on the desk in full view of the respondent so that he/she could immediately read what was required. So generally this section of the questionnaire was not a question and answer session but more of a cooperative endeavour. This worked well, the onus resting with the interviewer to ensure that the required information was collected for each work period. Second, the questions were repetitive for each work period so, even if the respondent did not immediately perceive what was required, this was quickly learnt and there was little need to actually ask the questions in a formal way. Finally, after a number of interviews had been undertaken, word began to get around the Polytechnic as to quite what information we were after and a number of people came 'prepared', usually with the list of their previous jobs and the dates written down on a bit of paper. None of these points undermined the quality of the data collected, in fact just the opposite is the case. A full and detailed Chronology was constructed for the vast majority of people interviewed. The only qualifications to this claim lie in the area of work evolving within a job and certain individual work histories, the topics of the next two subsections.

b) The construction of the Work History Chronology assumed that there was some identifiable point in time when the nature of work changed. This applied without difficulty in the majority of cases especially when people changed their employer.

However, it breaks down in those instances where work evolves imperceptibly over a long period of time. This was most noticeable amongst people who had held a job with the same title in the same workplace for many years. The work of a maintenance electrician in a car factory, for example, was no doubt different in 1950 to the work in 1980, yet the status remains constant and to the individual involved the changes were an ongoing development. A further problem here was a change in the job without any formal change in the nature of the work done. The clearest example of this was in the period immediately prior to either plant closure or large scale redundancy. Here the individual was still ostensibly doing the same job but the 'atmosphere' (a term often used by our respondents) had changed and the meaning of work was coloured by this. It has to be recognised that such subtle developments in an individual's work biography cannot be captured by the methodology employed in this project. The structured questionnaire used here has many advantages in terms of precision, comprehensiveness and comparability. But it does involve the loss of certain nuances which would be picked up by an oral history open ended list of topics to be covered (see Thompson, P. 1978).

c) The methodology employed in this project depended upon respondents being able to remember what work they had done in their lives, the sequence of work periods and when each began and ended. In general we experienced little difficulty from failures of memory apart from the month in which work periods began and ended. This was not because of psychological factors but because we were asking about events in people's lives which had been significant to them. The birth of a child is not easily forgotten! For the most part a change in work is similarly a significant event which sticks in the individual's mind. Yet this in turn rests on the individual having a fairly limited number of jobs and/or intervals between jobs. In two cases the person concerned had had just over twenty jobs and in both instances we were able to conduct the interview as planned. But in some other cases the sheer number of jobs and the character of the individual's work biography at significant periods of their life meant that the methodology broke down. This occurred in differing degrees. In one instance, for a brief period of time the person concerned had had a rapid succession of different jobs. In another instance, for a long period of time the individual had worked more or less continuously but casually for a wide range of employers. But in two other instances the person had had a large number of jobs stretching over a lengthy period of time, one person thought about 35, the other between 50 and 60! In these cases the work history methodology being described here had to be amended. This was not simply because of the practical difficulty of conducting interviews which, if the respondent had been able to recall the information, would have taken an inordinate length of time to conduct. Nor is it simply a question of the failure of memory. The people concerned could not recall all the jobs they had held, their sequence, or when they held them, because these were not significant jobs to them. At these points in their lives they were 'floating' members of the labour force for a variety of reasons which are not significant here but which made sense in each of the cases. Again, it has to be recognised that any research instrument has its limitations and that the one employed in this project was geared to people who had had a limited number of fairly stable work periods over the course of their lives. The fact is that this applied to the great majority of our respondents but in the instances mentioned adaptions were necessary. In the case of the person with a brief period of rapid movement between jobs, stage one details regarding occupation and industry were taken for the jobs remembered. No attempt to ask stage two questions for these jobs was made, neither is their sequence necessarily correct. But apart from this period the rest of the interview was 'as normal'. In the case of the person who worked casually, this period of life was treated as one, that is, simply as a period of casual employment. Again the rest of the work history was ascertained in the usual way. This was also done in cases where the person had emigrated to one of the dominions, worked casually for a while and then returned. However, in the cases of the two people with multiple jobs, stage one of the work history broke down, making stage two impossible. In these two instances the interview had to be quickly amended to become more of an unstructured attempt to get basic information. In both cases the range of different employment experience done by the individual was mapped out with the cooperation of the person concerned and then the comparison questions applied to these broader categories of work rather than to specific jobs.

d) A number of points are pertinent to the question of allocating each work period to the categories of 'work' listed above. Of course the categories are not necessarily mutually exclusive, a person can engage in voluntary work whilst being in employment and domestic work still needs doing whether a person is in employment or not. In all cases, paid employment was deemed to take precedence over any other kind of activity. This applied however 'small' the job in terms of the hours involved. So, a person employed for four hours per week as a part-time cleaner, whilst spending much more time working at home unpaid, is treated here as in paid employment. The reason for this was that, however small the job, it was hoped to compare people's evaluations and relate them to the characteristics of the job. However, it is worth noting that our respondents took the same view and even when working a few hours per week treated this as 'their job' with no prompting from the interviewer. A second issue here is the decision as to which of the categories applied. The major problem area lies in deciding whether a person, again usually a woman because of the system of benefit payment, is 'unemployed' or engaged in domestic work. The solution to this adopted in this project was to rely on the respondent's self definition. Thus when moving to a new work period the individual was asked to select what best described what was done next from a card listing, in straightforward language, the categories described above.

e) Information was obtained to establish the individual's occupation and industry throughout their working life. As is well known there are literally thousands of occupational titles and branches of industries, so some classificatory scheme was necessary to handle the enormous potential range of possibilities. The project adopted the Labour Force Survey's classificatory scheme as this permitted comparison with other data and the use of occupational data to establish people's place in a scheme of Socio-Economic Groups (SEGS). Any scheme which attempts to capture the wide variation in people's jobs leads to difficulties with the marginal or ambiguous case. The LFS scheme is a relatively abbreviated set of about 350 different occupational categories and 300 industries so at times it was something of a struggle to discover the appropriate code which accurately encompassed the information we had. Nevertheless these were difficulties at the margin rather than any systematic failure of the scheme to adequately reflect the data available. In only one area did we have any serious concern. This was in transforming occupations into SEGS using the OPCS **Classification of Occupations 1980** Appendix B in which the occupational categories we employed are allocated to the Census SEGS. On occasion we found that the information available to us clashed with the OPCS categorisation. This was especially on the borderline between skilled and semi-skilled manual work. We had a small number of cases in which the respondent had described the job to us as clearly semi-skilled, no difficulty was experienced in finding a descriptive occupational category for it but, then, in the OPCS scheme, this was classed as 'skilled manual'. Of course, what lies behind this is the social process by which jobs become categorised as 'skilled' or otherwise (Phillips A. and Taylor B. 1986, Cockburn C 1986). But from our point of view a methodological decision was required as to whether to blindly follow the OPCS scheme in the interest of consistency and comparability, or to be true to the information we had ourselves. The latter option was taken.

Stage two.

To move on to stage two of the work history. The reader will remember that the outline of the individual's work had by this stage been constructed and was 'visible' on the Chronology. For each work period, stage two asked a series of further questions, the replies being recorded on the questionnaire, a separate copy of the schedule of questions being used for each work period. Different questions were appropriate for paid and unpaid work, the latter including military service. As far as possible the replies to questions were precoded for ease of data entering and analysis, mainly requiring a 'factual' answer or a 'yes/no'. This proved successful. Some questions, however, were left open-ended and the coding scheme developed during the process of entering the data onto the computer. The range of information collected at this stage of the interview is indicated below together with comment on the data where necessary.

Stage two, data collected for work periods of paid employment.

1) How people obtained a job, distinguishing between those moving to a new employer and those whose work changed whilst with the same employer.
There is an ambiguity in our data here as between how a person first heard of the job and how they actually obtained it. For example, a person might hear from a friend that there are jobs at X but actually get the job by approaching X's themselves. I am afraid that our data contain this conflation of different stages in the job search process.

2) Training and post-school/college education. People were asked whether they were apprenticed, completed an apprenticeship, whether off-the-job training was provided by their employer and whether they themselves had undertaken any part-time education. Any qualifications obtained were noted. As all jobs require *some* training, even five minutes to show someone where equipment is kept, the criterion of 'off-the-job' was adopted to indicate a period of time when someone was actually being trained rather than doing a job.
A note is appropriate here on 'apprenticeship'. This question was asked only of those under twenty-one at the beginning of the work period. In planning the questionnaire it had been anticipated that how people would define 'apprenticeship' would vary and yet there seemed no straightforward way of providing a common and easily understandable definition. So this was left for the respondents to decide. Whilst most people who claimed to have been apprentices were provided a craft training, off the job education and certification and, in most cases, initiation to a trade union, there were some departures from this. On the one hand a few people equated apprenticeship with any training provided to a young worker. So one person claimed to have completed an apprenticeship as an office worker in a job lasting under one year. On the other hand, a number of people who were young workers during the Second World War were very aware that the war had prevented them from pursuing a 'proper' apprenticeship with the appropriate qualifications, even though they were trained and went on later in their trade. So, again, there is an ambiguity in the data here even though it reflects an ambiguity in the subject itself.

3) Whether people had a second job besides their main one.
There has been considerable debate in the literature on the incidence and significance of the informal economy. (E.g. South, N. 1982; Pahl, R.E. 1980, 1984). One aspect of this is people having second jobs in addition to their main one and it seemed interesting to ask this in the questionnaire, to see if there were links between

'moonlighting' and stages in the life cycle or the onset of economic recession. In some cases, holding a second job is frowned upon by employers so, when this question was asked for the current work period, it was prefaced by a further assurance of confidentiality. In spite of this it is clearly a crude instrument to explore what can be a very sensitive area. However, one interesting point which arose here was that several women workers defined their housework as their second job. At the data entering stage we assigned this response a separate code.

4) Number of workers at the workplace.
We did not expect people to remember this exactly, if indeed they knew in the first instance.[4] But from people's response ('thousands' or 'only a few') we could estimate which of a range of size bands the workplace fitted. Some ambiguity arose from the term 'workplace', especially where the person was working outside but based at a depot of some kind. Here the prompt from the interviewer was to ascertain the size of the significant working group from the respondent. So in the case of a lorry driver his depot was identified by him as the workplace whilst a painter would identify the gang he worked with as the significant group. In addition, we used a separate code for people in paid employment working at or from home.

5) Definition of a job as 'skilled' or not.
One of the dimensions on which it was hoped to compare different work experiences was the varying degree of skill associated with work in different contexts. Hence some estimation of the skill content of jobs was required. Now there are various ways of attempting to gauge the degree of skill required in a given job (e.g. Blackburn R.M. and Mann M. 1979, Chaps 2 & 3) but it was clearly not possible to apply these in retrospect and for the number of cases per respondent that we were covering in this project. Hence a very simple estimation of skill was required and to do this it was decided to use the respondent's estimation of whether the job was classified as skilled by the employer and by the respondent. With respect to the latter, a follow up open-ended question was asked to try to establish why the respondent took this view. These open-ended questions produced some very patchy responses. Some people would give a very detailed answer, others a perfunctory one. It was decided not to attempt to code and analyse this material as the quality of response often depended as much on the length, or anticipated length, of the interview. If the person had a lot of work periods to cover in stage two then the interviewer was aware of the necessity to press on, as indeed the respondent often was. In these cases there was little time to linger over the open-ended questions and tease out the details of what people meant by skill. Once again this involves recognising the limits of the research technique employed. The open-ended questions were an attempt to capture detail but in some cases this did not work.

6) Whether the person supervised others and was closely supervised. Again follow up open ended questions were asked but the points mentioned immediately above apply even more here. Often very stilted replies were received such as 'by the foreman' so that this data has not been utilised further.

7) Earnings.
There are numerous problems here. For a start, the self-employed, monthly salary earners and weekly wage earners needed to be sorted out as experience has shown that the first two tend to remember most accurately their gross annual earnings, whilst the latter remember their weekly net pay. This was achieved by using separate question formats. Second, a person's earnings in a job can clearly change over time. It would have been academically interesting to inquire about earnings when a person began a job and when he/she left so that comparisons with the next work period could be made.

But this would have stretched not only the memory but also the patience of the respondents too far. So earnings at the time the person left a given job were asked for. Again earnings can not only vary over time but also, especially amongst the weekly paid, from time to time. So we asked for an 'average' figure recognising that this can sometimes be misleading. Thus interviewers were aware not to average widely differing extremes of a range. Finally, this question did strain the memory of the respondents. Our impression at the interview stage, and at the data entering stage when the figure given was converted into 1985 value, was that the information given had severe deficiencies in some cases. This will be examined further in the analysis of 'best' and 'worst' jobs for money where further difficulties will be noted.

8) Whether the individual regularly worked overtime or worked shifts outside normal working hours.

9) Trade unionism.
Individuals were asked if they were a trade union member and if they were not why this was so. To ascertain an indicator of activism, members were asked whether they held office in the union and how often they attended branch meetings.

10) Whether the individual enjoyed the job 'on the whole' or not, and what they enjoyed and disliked about the job.
The job estimations would ask people for their overall assessments of their best and worst jobs so it was necessary to include some detail here about what the person had enjoyed and disliked about each job. These questions were rather crude but did allow us to obtain a brief summary of what was essentially liked and disliked about a particular job. Here our open ended questions worked well. A complex coding scheme was developed at the data entering stage.

11) Reason for leaving the job.

12) Spouse's economic status and occupation if applicable.
Ideally we would have liked this information about spouse's economic status for the whole of each respondent's work history. The reason for this is the speculation that one partner's work is very much interconnected with his or her opposite number. For example, it is clear that wives of unemployed men are much less likely to be in employment themselves than wives of employed men. (Hakim, C. 1982 McKee, L. & Bell, C. 1985). In a number of our respondents' cases it was clear that the person's economic biography was connected to his or her spouse's. For example one woman had joined her husband in buying a general store, both of them giving up 'good' jobs. But after some years the store had become unviable through the opening of an adjacent hypermarket and both partners had had to take what they regarded as much more menial employment. But in the context of this study the collection of partner's work history (not to mention other members' of the household) was simply impractical. Clearly the length of any interview depended on the number of work periods but on average they lasted an hour and a half. To have added this would have strained the feasible time unrealistically. In addition to this, there are real difficulties about ascertaining data about one person from the recollections of another. Hence the compromise was simply to ask for this information for the present.

Stage two, data collected for work periods not in paid employment. The questions asked here largely depended on which of the non-paid economic categories applied, however in each case people were asked why they were not in employment at the time and why they moved to the next work period. In the cases of those who were long term sick or in training, education or voluntary work, people were asked to give the basic details of what they were doing at the time. Those in military service were asked whether they received training leading to qualifications Those doing unpaid domestic tasks were asked if they had undertaken any further education at this time and, if so, if any qualifications were obtained.

A particular problem related to periods of domestic work. It was possible that in identifying herself (generally) as in domestic work a woman omitted some paid activity which she undertook at this time either overlooking it or disregarding it as 'just a little job'. Examples would be acting as agents for home selling operations such as Avon or Tupperware, outwork, or delivering 'free' newspapers or leaflets. We did not wish to miss such activities so people were asked:

You were mainly concerned with looking after the family at this time but did you do any work for which you were paid?

As mentioned above the presupposition of the project was that if paid work was done that should take precedence and the battery of questions for paid work, stage two should be asked. But this could be misleading if a housewife had had very spasmodic employment, for example, working as a shop assistant just before Christmas. Hence the interviewer was instructed to record any paid work undertaken and, if this was regular, to go back to the Chronology, revise it and treat the paid work as a new work period. But where the employment was spasmodic it was simply recorded and the work period remained as domestic work.

It is opportune here to mention the more general problem which lies behind this, how 'multiple' economic statuses were to be handled? People with several paid jobs are the most obvious example. The general solution to the problem was to regard the highest status job (to be judged pragmatically) as the main job and to regard further jobs as second to this.This broke down in one instance where the person's current job at the Polytechnic was, in fact, her second job and of lower status than her main job. To have followed the above procedure in this case would have meant, in effect, excluding the person from the sample. In this case the *ad hoc* procedure of making the person's main job and the Polytechnic job successive rather than concurrent was adopted.

'Best' and 'worst' jobs and their comparability with present job.

The interview had now reached the stage where we had a detailed Chronology of the person's work periods and further detail on each one, especially for paid employment. The final section of the questionnaire was designed to inquire about the person's present job compared with other work experience in the past. At an early stage in planning the research it had been thought that at this point the interview would switch to a less structured approach to allow people to talk more freely about their work biographies. This was decided against for two reasons. First, as the material required for the work history, stages one and two, became clear it was evident that time was of the essence and that a relaxed informal discussion at the end of a fairly intensive structured interview would be impractical. Second, at this time, by coincidence I happened to have the opportunity to inspect some transcripts of non-structured interviews done on another project. The length of these, the work necessary simply to transcribe them, and the enormous complexity of the analysis made me think again and seek a simpler way of addressing the problem.

The strategy adopted was to ask people to identify their 'best' and 'worst' jobs in various specific respects. Remember again that the Work History Chronology was available for the respondent to peruse during these questions. The person was then asked how his or her present job compared with the one identified, unless, of course, the best/worst job was the present job. The comparison was made in terms of whether the present job compared 'very well', 'well' or 'not much different' in the case of worst jobs and 'very badly', 'badly' or 'not much different' for best jobs. The aspects of jobs covered were financial rewards, skill required, degree of job autonomy, the services of a trade union and best/worst jobs 'all things considered'.

A number of problems were generated by these questions but as they influence the quality of the data available it is best to integrate them into the substantive discussion of the responses in chapters six to ten. However, a couple of further points are of note here. In the preceding paragraph the term 'job' has been used to describe a work period. But in view of the desire to incorporate non-paid work into the range of comparison it was felt necessary to prompt the respondent on this because in common sense parlance to speak of work is very often to restrict oneself to paid employment. This being the case, domestic work, for example, would be omitted from the comparisons and the possibility of such work being identified as 'best' or 'worst' would be excluded. Hence the section began with the following preamble:

I now want to ask some questions covering all your working life. We usually think of a 'job' as work for wages or earnings but in these questions I want you to regard any times when you were not working for wages or earnings as a 'job' just like the paid work you have done. Have I made myself clear on this point?

ENSURE THAT RESPONDENT IS CLEAR, USE RELEVANT EXAMPLES.

In fact this strategy was a failure. Although our impression was that people took our meaning at the beginning of the section, they rapidly 'reverted' to the colloquial sense of 'work' and 'job' and few people identified non-paid work in any respect. In cases where they did, we had a backup question asking them to make the comparison simply in terms of paid employment, so that in this respect we have full data. But in view of the interviewing experience it seemed pointless to include the few cases in the analysis, especially since even here the range of comparison was spasmodic, in that people included non-paid work in the earlier questions before forgetting this point later on. Hence in the data entering, identifications of non-paid 'best'/'worst' jobs were passed over in favour of the follow up paid job.

Finally, it was, of course, feasible that some of our respondents would not be 'eligible' for these questions for the simple reason that they had had only one work period since leaving school or college. To cover this possibility a series of questions was included as an alternative to the 'best'/'worst' section, asking people whether they had intended staying at the Polytechnic so long when they first came and what their future intentions were. In fact, none of our sample came into this category and these questions were not used. However, this possibility of noncomparability because of only one (or no) instance did also crop up in the trade union questions which simply meant recording the person as either never a trade union member or only a member in one job.

Notes.

1. I.e. whether they were cleaners, porters, caterers, protection officers or technicians.

2. See chapter 3.

3. The 'Other' category was included to incorporate periods of imprisonment as well as any activity not anticipated by the researcher. In fact, from what we were told none of our respondents had been imprisoned and we did not perceive any lack of validity here.

4. Hardly anyone, for example, could venture an estimate of the Polytechnic's workforce.

3 The workforce, sampling principles and some background characteristics of the sample

This chapter sets out to cover five areas. First, to give a brief outline of the structure of the workforce in terms of broad occupational groups, sex and full/part-time status. These three criteria were the main considerations used in drawing up the sample of individuals asked to be interviewed. The second task is to outline the *intended* sample. However, there is some disparity between this intention and the *actual* sample achieved. To account for this note has to be taken of the method of approach to individuals and how the actual sample was finally arrived at. This is covered in section three, followed by a sketch of the actual sample in section four. In the final section the occupational, sex and full/part time characteristics of the sample are supplemented by further background material with the intention of giving a general picture of certain features of the workforce. These include age, marital status, family situation, the economic status of the spouses of the sample members, their length of service at the Polytechnic, how they obtained employment there, and job movement within the Polytechnic.

The total workforce.

As explained in the Introduction, we were interested in interviewing people working as ancillaries and technical staff. The Polytechnic supplied a list of workers in these occupations indicating, for each person, their job title, whether they were male or female, and whether they were employed on a full or part-time basis. These items of information were necessary in order to draw up a meaningful sample of the workforce. It has already been noted that the labour force participation of men and women is rather different so that a balance of the two was required. As well as the difference between the continuity of participation over time, there is the related variation between men and womens' employment on a full or part-time basis. So this again needed to be included in the construction of the sample. Finally, the range of occupations covered by the phrase 'ancillary and technical workers' needed consideration. The major occupational groups here were people involved in cleaning, catering, various portering jobs, protecting the site and its occupants, and, on the technical side, those working as technicians in the Polytechnic's academic departments and those engaged in the maintenance of the campus. Unfortunately, the distinction between the two groups of technicians was not appreciated at the time the sample was drawn and in this chapter the two are treated together simply as 'technicians'.

The list of names supplied to me was current for the spring of 1985. At that time the Polytechnic employed 621 people in these ancillary and technical positions. These included 345 women and 276 men. 310 of those employed held part-time posts, 311 full-time. In terms of occupations the workforce was distributed as in table 3.1.

Table 3.1
Total workforce by broad occupational group.

Occupational group	Number	Percent of total
Cleaners	232	37
Porters	24	4
Caterers	98	16
Protection	28	5
Technicians	237	38
Others*	2	0
Total	621	100

* The two others were a gardener and a seamstress who fell outside the occupations investigated.

It is when these three considerations are put together that the main features of the workforce become clear. This is done in table 3.2. The workforce is clearly highly segregated in terms of these three characteristics. The vast majority of men (all but two) work on a full-time basis, whilst most of the women (308 of the 345) are employed part-time. Alongside this is the segregation of men and women in the different occupational groups. 216 of the cleaning staff of 232 are women, 94 of the 98 caterers. All the protection officers and portering staff are men. The majority of technicians are men although there is a significant minority of women here, most of them full-time. But, on the whole, the workforce is composed of part-time women cleaners and caterers, full-time men protection officers, porters and technicians, with smaller groups of male cleaners and female technicians, and a smattering of others.

Table 3.2
Total workforce by occupational group,
sex and full/part-time status.

Occupational group	Men F-t	Men P-t	Women F-t	Women P-t	Total
Cleaners	15	1	2	214	232
Porters	24	0	0	0	24
Caterers	4	1	5	88	98
Protection	28	0	0	0	28
Technicians	202	0	30	5	237
Others	1	0	0	1	2
Total	274	2	37	308	621

The intended sample.

Having identified the main outlines of the structure of the workforce, a sample had to be drawn to allow for representation of these various features. In addition, a further point had to be taken into account. The various job titles were grouped into five occupational groups. This assumes a degree of homogeneity amongst the jobs put together. In some cases this was fairly self-evident, especially amongst the women cleaners and the protection officers. In others it was not so and the sample would have to reflect the heterogeneity of jobs grouped together.

Bearing these points in mind it was clear that a simple random sample would not be suitable. Resources would allow up to one hundred interviews, that is, a roughly one in six sample of the total workforce. But to simply take every sixth name on the list of employees would swamp the smaller groups such as the protection officers, porters and women technicians. Thus it was decided to use different sampling fractions for the five occupational groups and then, within these, to take account of the sex and full/part-time ratios as well as the degree of homogeneity of the jobs covered.

The sampling fractions used for the occupational groups were:

Cleaners	25	i.e. approx. 1 in 9
Porters	12	i.e. 1 in 2
Protection	13	i.e. approx. 1 in 2
Caterers	20	i.e. approx. 1 in 5
Technicians	30	i.e. approx. 1 in 8

To show how the other considerations mentioned above were taken into account, each of these occupational groups can be taken in turn.

The **cleaners** broke down as follows:

214	part-time women
1	part-time man
2	full-time women
15	full-time men

Given the overall size of this group the single part-time male and the two full-time women seemed anomalous and were thus omitted from the sample. To the best of my knowledge, the large group of women part-timers were reasonably homogeneous in terms of the work they performed, so that a smaller fraction of their total could suffice. On the other hand, whilst the full-time men were again apparently homogeneous, a rather large fraction was necessary to give a subgroup large enough to be meaningful. On this basis the cleaning group was subdivided into target numbers of 19 part-time women and 6 full-time men.

The **porters** were all full-time men. The group included a range of portering, driving and attendant jobs but the large sampling fraction was quite adequate to take this into account.

The **caterers** were made up of:

88	part-time women
1	part-time man
5	full-time women
4	full-time men

Here the decision was difficult, the single part-time man could be set aside as anomalous and the full-timers of both sexes were very small. In addition a relatively large number of part-time women were required here to keep their overall number representative. So it was decided to include only part-time women workers from the catering staff. As it turned out some full-timers appeared in the actual sample because of staff changes between the time my staff list was drawn up and our actual interviews in the autumn.

The **protection** staff were all male full-timers and the fraction aimed for would incorporate the variation between jobs.

The **technicians** were composed of:

5	part-time women
0	part-time men
30	full-time women
202	full-time men

Here the part-timers were clearly very few and could be set aside. In addition the problem of heterogeneity was considerable here. The 'technicians' included the Polytechnic's maintenance staff and technical workers in academic departments. The former fell into five different trade groups and the latter covered the range of disciplines offered by the Polytechnic. Furthermore, the technicians are graded according to seniority. Given the numbers available, each of these sources of variation could not be systematically taken into account. The main point seemed to be to make the overall group large enough to attempt to cover the range of jobs and to include a relatively large number of women full-timers as this occupational group was just about the only area where they were to be found in any number. Thus, as noted above, a sampling fraction of one in eight was used to give an intended sample of thirty. These were to be made up of twenty men and ten women, all full-time.

Overall, then, the intended sample was to be structured as in table 3.3.

Table 3.3
Intended sample by occupation, sex and full/part time status.

Occupation	Men		Women		Total
	Full	Part	Full	Part	
Cleaners	6	0	0	19	25
Porters	12	0	0	0	12
Caterers	0	0	0	20	20
Protection	13	0	0	0	13
Technicians	20	0	10	0	30
Totals	51	0	10	39	100
	51		49		100

Achieving the sample.

Having arrived at the numbers required amongst the various groups, it was next necessary to make certain decisions as to how to achieve the desired sample. As mentioned above, the total overall sample was intended to be 100 people. As the interviewing was to take place in 'works time' and had the support of both management and trade unions, a high response rate was anticipated and so only a small allowance was thought to be necessary to achieve 100 interviews. 25% was added to the target number of the following groups to arrive at the number of people in each who would be asked to participate.

Group	Target	Number invited
Cleaners, women, p/t	19	24
Cleaners, men, f/t	6	8
Porters, men, f/t	12	16
Caterers, women, p/t	20	25
Protection, men, f/t	13	16
Technicians, men, f/t	20	25
Technicians, women, f/t	10	13
Totals	100	127

Having determined the numbers to be targeted in terms of the sampling criteria selected, each person on the total list of staff was allocated a number and the sample drawn by either using random number tables, in the case of the small subgroups, or systematic random sampling in the case of subgroups with large numbers.

The 127 people selected were approached by a letter sent to them in late August 1985. On one side of the letter they were asked to participate in the project, informed that although the project was supported by the Polytechnic management and trade unions it was an independent piece of academic research, assured that their names had been picked at random and given an absolute guarantee of confidentiality, and asked to return a reply slip in an enclosed pre-paid addressed envelope. On the other side of the letter a brief account of the main objectives of the research project was provided for information. The letter was distributed through the Polytechnic's internal postal system.

Over the succeeding weeks the reply slips arrived back with me. It quickly became apparent that my anticipation of a high positive response rate had been over optimistic. At this stage, 49 people agreed to be interviewed, 30 people declined, 9 had left the Polytechnic and thirty-nine people did not initially return the reply slip. It seemed that the pattern of response was worth analysing in terms of the three sampling considerations of occupation, sex and full/part time status to see if there were any patterns of variation which would be instructive in attempting to boost the response rate.

To begin with the initial response rates by sex in table 3.4.

Table 3.4
Initial response by sex.

Sex	Yes	No	No reply	Left	Total
Number of men	22	16	26	1	65
% of men	34	25	40	2	101
Number of women	27	14	13	8	62
% of women	44	23	21	13	101
Total number	49	30	39	9	127
% of total	39	24	31	7	101

Here, proportionately more women than men had agreed to be interviewed, there was little difference in their rate of decline but considerable variation in the 'non repliers' with men not replying far more than women. Also, more women than men had left the Polytechnic.

Table 3.5 summarises the initial response by full and part-time status.

Table 3.5
Initial response by full/part time status.

Status	Yes	No	No reply	Left	Total
Number of f/t	29	19	27	3	78
% of f/t	37	24	35	4	100
Number of p/t	20	11	12	6	49
% of p/t	41	22	25	12	100
Total number	49	30	39	9	127
% of total	39	24	31	7	101

There is little in the way of variation here apart from a rather larger proportion of full-timers not replying and a greater proportion of part-timers leaving the Polytechnic.

Table 3.6 presents the picture of initial response by the five occupational groups used in the sample structuring.

Table 3.6
Initial response by occupational group.

Occupation	Yes	No	No reply	Left	Total
Number of cleaners	14	8	8	2	32
% of cleaners	44	25	25	6	100
Number of porters	3	4	9	0	16
% of porters	19	25	56	0	100
Number of caterers	10	5	6	4	25
% of caterers	40	20	24	16	100
Number of protection	7	3	6	0	16
% of protection	44	19	37	0	100
Number of technicians	15	9	11	3	38
% of technicians	39	24	29	8	100
Total number	49	30	39	9	127
% of total	39	24	31	7	101

Here again there is no marked difference in response, with the clear exception of the porters, with a very low proportion agreeing to be interviewed and a high rate of non reply.

From this exercise, it appeared that the next step should focus on the 'non repliers', the 39 people who had not agreed but had not actually said no. They constituted a fairly high proportion of the sample, so that a positive response would boost the numbers. Clearly a different tack was necessary than further postal communication. It was decided to approach each non replying individual personally to ask if they would participate. It was felt that this would have the additional advantage of revealing why people were not willing to participate if the non repliers in fact turned out to be 'decliners'. Apart from the overall sex difference, of which I could make little sense, the only marked exception to the overall pattern was that of the porters. Indeed, closer inspection of the responses here showed that the three individuals who had agreed were attendants of various kinds and that, in fact, none of the portering staff as such had agreed to be interviewed. Clearly this group required a rather different approach.

The method adopted was to approach a key individual. This was the senior Transport and General Workers Union shop steward who had agreed to the project and who was a porter himself, although not in the sample. By chance it was break time and I was taken into the porter's rest room and had a 15 minute discussion with five of the porters. I am afraid that from the point of view of the research I made no progress, there was a general unwillingness to participate and the men were not to be moved. This, of course, they were perfectly entitled to do, although as a researcher the situation was puzzling as, amongst people who had agreed, there was a remarkable openness to the interviewer and, when I approached the 'non repliers', those who did not want to participate did not show the hostility to the prospect of interview that the porters I briefly met did. So the porters remain a puzzle. From bits of observation I can offer some speculations about their refusal to participate but I must stress that these are speculative. It would have been of interest to see if the porters' pattern of work history was very different from the other people we spoke to! First, the reason offered by their shop-steward, which was that the men had mostly worked in the past in factory environments and whenever they had been 'surveyed' it had worked to their disadvantage. This might well be the case but as a great many of the other people who agreed to be interviewed had also worked in such environments, this in itself was hardly convincing. Again, in my observation, the porters seemed a cohesive group of workers who often worked as a team and had their own rest room where they took their breaks together. So the unanimity of their response could be connected to the cohesiveness as a group. But other workers at the Polytechnic also seemed to have such coherence, the protection men and the maintenance technicians especially. Finally, the status of the porters in the Polytechnic may be involved. A few years ago the tasks of portering and protection were split from each other and a separate body of protection officers created. These were given staff status and it may be that some resentment of this persists amongst those remaining as porters. Together these three factors may account for the unwillingness of the porters as a body to cooperate with the research project.

As mentioned above, with the exception of the nine porters, the thirty others who had not replied were approached individually and asked to participate. Eighteen agreed to be interviewed. Of the twelve who declined, a few said that their work did not leave them any time to spare but, in the main, a general lack of interest in 'that sort of thing' was offered as the reason.

The actual sample.

The final response pattern was as shown in table 3.7.

Table 3.7
The sample by occupation,
sex and full/part-time status.

Occupation	Men		Women		Total
	F	P	F	P	
Cleaners	6	0	0	13	19
Porters	3	0	0	0	3
Caterers	0	0	0	12	12
Protection	10	0	0	0	10
Technicians	15	0	8	0	23
Total	34	0	8	25	67

This can be broken down to examine how the sample compares to the intended sample and the actual workforce in terms of the three sampling criteria of occupation, sex and full/part-time status.

To begin with occupation, the relative proportions of the five occupational groups in the sample, the intended sample and the workforce are as shown in table 3.8. The actual sample succeeded in achieving the intended sample to a reasonable extent. In proportional terms, rather more cleaners, protection officers and technicians were included than had been intended and rather less caterers. But the difference between the intended sample and that achieved was not large in these four groups.

The porters are the exception, the intention was that 12% of the sample would be porters whilst the actual sample included only 4% for the reasons offered above.

Table 3.8
Percentages in sample, intended sample and workforce by occupation.

Occupation	% in sample	% in intended	% in workforce
Cleaners	28	25	37
Porters	4	12	4
Caterers	18	20	16
Protection	15	13	5
Technicians	34	30	38
Total	99	100	100

By comparison with the workforce, the actual sample under-represents the cleaners significantly and slightly under-represents technicians. Caterers are slightly over-represented and protection officers markedly so. As things turned out porters are represented in the sample in their relative proportion in the workforce.

Table 3.9 gives the sex distribution of the sample as compared to the intended sample and the workforce.

Table 3.9
Percentages in sample, intended sample and workforce by sex.

Sex	% in sample	% in intended	% in workforce
Men	51	51	44
Women	49	49	56
Total	100	100	100

The actual sample thus achieved the intention of over-representing men. It will be remembered that this was to incorporate larger groups of protection officers and porters than a simple random sample of the workforce would have achieved. This was to be attained at the expense of the under-representation of women cleaners. In fact, this intention was arrived at in skewed fashion, the men working as protection officers are over-represented and the porters under-represented.

The full and part-time status of the sample, intended sample and the workforce are portrayed in table 3.10. Again the actual sample more or less achieved the intended result, in this case in the way intended, that is, by over-representing male full-time cleaners and full-time female technicians.

Table 3.10
Percentages in sample, intended sample
and workforce by full/part-time status.

F/P Status	% in sample	% in intended	% in workforce
Full	63	61	50
Part	37	39	50
Total	100	100	100

It is appropriate here to note the number of hours worked by the two groups of part-time workers. Eleven of the thirteen women cleaners reported a working week of eighteen hours, one a week of twenty hours and one twenty-one hours. The caterers were rather more complicated. On the information supplied to us for sampling purposes all the caterers selected in the sample were part-time. However, in the event, three worked a twelve hour week, three a twenty-five hour week, two reported their hours as thirty-three hours, two thirty-five and one forty. (One case is missing). Those working over thirty-three hours all designated themselves as full-time, of the two working thirty-three hours, one described herself as part-time, the other full-time.

Further characteristics of the sample.

The composition of the sample in terms of sex, broad occupational group and full/part-time status has been indicated. In this section we will say a little more on these matters and give further detail on the characteristics of the people in the sample. At this point this will be confined to age, marital and familial situation, length of service at the Polytechnic and how people obtained their employment there. In later chapters other information will be deployed in relation to the sample's work history and their estimations of 'best' and 'worst' jobs.

To turn first to the age structure of the sample. The average age of the sixty-seven people was 42.4 years. Table 3.11 shows how individual's ages were distributed between various cohorts and compares this with the national picture for the employed population as displayed in the 1981 Census. This shows that the sample's age distribution is rather distinctive compared to the national picture. The male employees in the sample are concentrated in the older age cohorts. The Polytechnic employs very few young men under 30, rather less than the average of men in their thirties and markedly more men over 40, especially over 60. Amongst women the pattern is rather different. Again there is an absence of teenage workers and rather fewer women in their twenties than the national picture but then a considerable bunching of women between 30 and 49 before the sample figures drop below the national for older women.

As well as these differences between the sample and national statistics it is clear from table 3.11 that the age structure of men and women in the sample also differs. The average age of men was 47, of women 36. Their distribution in age cohorts is shown in table 3.12.

Table 3.11
Age cohorts as percent of male/female employed population. G.B. 1981 and the sample.

Cohort	Men		Women	
	G.B.	Sample	G.B.	Sample
16-19	7	0	9	3
20-29	22	9	23	21
30-39	24	21	22	33
40-49	20	23	22	27
50-59	19	23	19	12
60-64	6	23	4	3
Rest	2	0	1	0
Totals	100	99	100	99

G.B. Source: OPCS Census 1981, Economic Activity G.B. Table 3.2

Table 3.12
Age by sex.

Range	Men	Women	Total
16-19	0	1	1
20-29	3	7	10
30-39	7	11	18
40-49	8	9	17
50-59	8	4	12
60-65	8	1	9
Totals	34	33	67

Table 3.13
Mean average age by sex and occupational group.

Occ. Group	Men	Women
Acad. tech.	47	30
Mainten. tech.	42	
Protection	48	
Cleaners	56	43
Caterers		37
Attendants	39	

There is also variation in the age structure of the sample in occupational terms as table 3.13 illustrates. Amongst the women two of the occupational groups have an average age in the thirties whilst the women cleaners are considerably younger than their male counterparts. The three porters have an average age of 39 and with this slight exception all the male groups have mean ages of over 40. Note the age disparity between the male academic technicians (47) and the women technicians (30).

Of the sixty-seven people in the sample, fifty-five were currently married, (28 women and 27 men) one man was divorced, one woman had in the past been a partner in a non-legalised marriage and ten (4 women and 6 men) had never been married. Seven of these last ten were aged 25 and under. Comparing these figures to the national picture,[1] a high proportion of the people in the sample were currently married. For men in the sample it is 79% whilst in Great Britain as a whole it is 69%. For women the sample percentage is 85%, the national figure 62%.

A high propensity of marriage holds across most of the occupational groups making up the sample but table 3.14 shows that there are some exceptions to this. Only half the female technicians were married. It will be remembered that these were also the youngest group in the sample, which perhaps accounts for this, although, as we shall see, a number of other points distinguish this group. It can also be noted that half the male cleaners were not married.

Table 3.14
Marital status by
sex and occupational group.

Sex and occupation	Married	Divorced	Separated	Single	Total
Male acad. tech.	10				10
Female acad. tech.	4		1	3	8
Maintenance tech	4			1	5
Protection	8			2	10
Male cleaners	3	1		2	6
Female cleaners	13				13
Caterers	11			1	12
Attendants	2			1	3
Totals	55	1	1	10	67

Turning next to the family situation of the sample, this is displayed in table 3.15 together with a percentage comparison with national figures. Nationally 43% of families are made up of parent(s) and their children. In the sample, 50% of the individuals lived in this type of family. But here the high proportion of the sample who are currently married is also worth remembering. None of the people in the sample formed lone parent families with their children whilst, in the national sample, 5% of families were of this type. Thus nationally 38% of families included married parents and children whilst in the sample the figure was 50%.

Although most of the thirty-four families with children included at least one child under 17 years of age (27), the average age of children living at home was 16. There was little variation in this figure in terms of either the sex of the individual or their full or part-time status. The average age of the men's children living at home was 16.6 years,

Table 3.15
Family type, the sample and G.B.

Family type	Number	Percent Sample	Percent G.B.
Married, child under 17 at home	27	40	30
Married, only children over 17 at home	7	10	8
Married, no children at home	21	31	27
Lone parent, child under 17 at home	0	0	5
Others*	12	18	31
Total	67	100	101

* For the sample the others category includes only those not currently married.

The G.B. figure includes 1 person families, 2 or more related adults, lone parents with children aged 17 and over and two or more families living as one.

G.B. Source: General Household Survey 1983 Table 3.5 Pp. 14.

women's 15.2. For full-time workers, the average age of children living at home was 16.2 years, part-time workers' children had an average age of 15.4 years. As this indicates, there is little variation in these respects as to family type. Of the sixteen men with children at home, twelve families included at least one child under 17, four included only children 17 and over. For women, the eighteen families included fifteen with a child under 17 and three with only children aged 17 and over.

There is some variation within the sample in terms of children's ages and family type in relation to the various occupational groups. The data is displayed in table 3.16. The sex/occupational groups tending to have children at home are the male technicians, maintenance technicians, female cleaners and caterers. Those tending not to have children at home are the female technicians, protection officers, male cleaners and attendants. This seems largely related to the age characteristic of these groups. The women technicians are relatively young and some are not married. The protection officers and male cleaners are relatively old. These points account for why they tend not to have children or not have children at home. On the other hand, the maintenance technicians, women cleaners and caterers are of an age when family formation is likely, note how the caterers are rather younger than the cleaners and how their children are correspondingly rather younger on average. However, one group does not fit this pattern. The male technicians are relatively old by comparison with the sample as a whole, with an average age of 47 years. Yet over half of them still have children at home and their average age is 13, the youngest average figure of any of the groups in the table. This might well be linked to the work history pattern of these men which we note in chapter five.

A large proportion of the spouses of the married workers in the sample were in paid employment, forty-one of the fifty-four.[2] Of the remaining thirteen, three were unemployed, all husbands, seven doing unpaid domestic work and three others occupying a non-wage capacity. All of these unwaged workers were women, with one exception, a husband receiving disability benefit.

Table 3.16
Family type and childrens' ages
by sex and occupation.

Sex and occupation	Age	Married, child under 17	Married, child 17 & >	Married, no children	Other	Children's average age	Total
Male acad. tech.	47	5	1	4	0	13	10
Female acad. tech.	30	1	0	3	4	16	8
Maint. tech.	42	3	1	0	1	15	5
Protection	48	2	2	4	2	23	10
Male cleaners	56	1	0	2	3	22	6
Female cleaners	43	6	1	6	0	17	13
Caterers	37	8	2	1	1	14	12
Attendants	39	1	0	1	1	14	3
Totals	42	27	7	21	12	16	67

Table 3.17
Socio-Economic Group of sample's spouses.

S.E.G.	No.	Percent	
2.1 Managers, large establishments	2	3.7	
2.2 Managers, small establishments	1	1.9	9.3
4 Employed professionals	2	3.7	
5.1 Non-manual ancillaries	8	14.8	
5.2 Non-manual foremen	1	1.9	31.6
6 Junior non-manual	5	9.3	
7 Personal service workers	3	5.6	
8 Manual foremen	1	1.9	
9 Skilled manual	8	14.8	35.3
10 Semi-skilled manual	7	13.0	
11 Unskilled manual	3	5.6	
Domestic work	7	13.0	
Other non-waged	3	5.6	24.2
Unemployed	3	5.6	
Total	54	100.4	100.4

12 members of the sample were not married. 1 case is missing.

The Socio-Economic Group of the sample's spouses is shown in table 3.17. At the level of the sample as a whole, the occupations of spouses are spread quite evenly through the class system with slight bunchings in the non-manual ancillary, skilled manual and semi-skilled manual groups. However, when we break this down by sex and the occupational group of the sample member, quite noticeable differences emerge. To make this more manageable, the Socio-Economic Groups listed in table 3.17 have been aggregated into three broader categories; 'higher white collar', 'lower white collar' and 'manual' workers. The first contains S.E.G.'s 2.1, 2.2 and 4 in table 3.17, that is, managerial and professional occupations. The second contains S.E.G.'s 5.1, 5.2, 6 and 7, that is, people working as ancillaries to higher white collar workers, non-manual supervisors, routine clerical and personal service occupations. The third group contains the manual worker S.E.G.'s 8, 9, 10 and 11. In addition, we continue to deploy a category for unwaged work (combining domestic work and the 'other' category) and one for unemployed people.

Table 3.18 displays the occupational position of the spouses of sample members broken down by sex, that is whether the spouse is a husband or a wife.

Table 3.18
Occupational group of sample members' spouses by sex.

Occupational group	Husbands	Wives	Total
Higher white collar	4	1	5
Lower white collar	3	14	17
Manual	16	3	19
Unwaged	1	9	10
Unemployed	3	0	3
Total	27	27	54

12 members of the sample were not married, 1 case is missing.

Table 3.18 shows that the great majority of husbands were to be found in manual occupations, the majority of wives in white collar occupations. However the latter tended to be in the lower white collar group, only one wife was to be found in the managerial and professional group. Those husbands who were to be found in white collar occupations were split between the two, with four in the higher, and three in the lower categories. What this reflects is the well known differentiation of occupations as between men and women. If we take this a step further, in table 3.19, we see that there is also a quite strong link between the sample members' occupation and that of his/her spouse. All the spouses of the technician groups and the protection officers were to be found in the white collar groups, largely in the lower white collar category and these were largely the wives of male Polytechnic workers. Most of the spouses of the cleaners (male and female), caterers and attendants were to be found in the manual group. There were some exceptions here. Two women cleaners and one caterer had a husband in a higher white collar occupation, one cleaner and one attendant a spouse in the lower white collar group.

We now turn from the family situation of the sample to their length of service at the Polytechnic. At this point we should note the distinction between length of service with the *employer* and length of service in a particular *job*. Clearly people can change their

Table 3.19
Occupational group of sample members' spouses.
By sex and occupation of sample member.

Sex and occupational group	Higher white collar	Lower white collar	Manual	Unwaged	Unemployed	Total
Male acad. tech.	0	6	0	4	0	10
Female acad. tech.	1	2	0	0	1	4
Maintenance	1	1	0	2	0	4
Protection	0	6	0	2	0	8
Male cleaners	0	0	2	1	0	3
Female cleaners	2	1	8	1	1	13
Caterers	1	0	8	0	1	10
Attendants	0	1	1	0	0	2
Total	5	17	19	10	3	54

12 members of the sample were not married, 1 case is missing.

jobs whilst remaining with the same employer and we will see shortly that this applied in twenty-six of the sixty-five cases for which we had data. But here we are concerned to give an indication of the length of service with the Polytechnic as an employer so that the following relates to the time when individuals *first* took up employment there.

For the sample as a whole the average number of years in the Polytechnic's employment was 6.9. There was a noticeable difference here as between men and women. Male workers had an average length of service of 8.4 years, women 5.4 years. However although there was a general tendency for men's stability of employment at the Polytechnic to be higher than women's, there are also considerable differences as between occupations as table 3.20 indicates. Within the male workers, the academic technicians (especially), maintenance technicians and cleaners all had average lengths of service above the average but the protection officers and attendants were below average. Again, within the women workers there is a difference between the cleaners and technicians on the one hand and the caterers on the other. Within occupations the variation between the men and women technicians is particularly noticeable, as is that between the male and female cleaners. However we can note that the latter is not simply to do with the full-time status of the male cleaners and the part-time women. Most of the caterers were part-time but had a quite long length of service, whilst all the women technicians were full-time but had a relatively short length of service.

Nevertheless, although these variations are of note, the main point here is that most of the sample had worked at the Polytechnic for a good number of years, especially men and especially those in technician and cleaning occupations.

There is also some variation in the age at which people began their Polytechnic employment but the average is the first point to note. For the sample as a whole it was 36.2 years, varying between 39.3 for men and 32.9 for women. In general then people began their Polytechnic employment in their thirties. When we look at this in occupational terms we see the variations displayed in table 3.21.

Table 3.20
Number of years at Polytechnic.
By sex and occupation.

Sex and occupation	Number of years
Male acad. tech.	13.0
Female acad. tech.	4.6
Maintenance	7.4
Protection	5.0
Male cleaners	9.7
Female cleaners	4.6
Caterers	6.8
Attendants	4.0
All	6.9

Table 3.21
Age when employed at Polytechnic.
By sex and occupation.

Sex and occupation	Age when began
Male acad. tech.	34.6
Female acad. tech.	25.4
Maintenance	34.8
Protection	43.2
Male cleaners	46.5
Female cleaners	39.8
Caterers	30.5
Attendants	35.3
All	36.2

The occupational groups above the sample average age were the male cleaners, protection officers and female cleaners. Those below the average, the caterers and women technicians (especially), with the attendants and male technicians being close to the average.

The last point we will consider in this sketch of background characteristics of the sample is the means by which people obtained their initial employment at the Polytechnic and their present jobs. Table 3.22 shows how people first obtained employment at the Polytechnic. What is immediately noticeable is the small proportion of people obtaining their employment through the 'formal' means of the job centre and newspaper advertisements. Only a quarter of the sample found their job this way and in only one case was this through the job centre. On the other hand, sixty-five percent of the sample found their first job at the Polytechnic by more 'informal' means, especially hearing of a job from friends or relatives and, within these, particularly from someone already employed at the Polytechnic.

Table 3.22
Means of obtaining first job at the Polytechnic.

	No.	%
Job centre	1	1.5
Newspaper advertisement	16	24.6
Direct approach to employer	10	15.4
Approached by employer	3	4.6
Friend/relative employed at Poly	24	36.9
Friend/relative not employed at Poly	5	7.7
Transfer from Coventry C.C.	6	9.2
Total	65	100

Two cases are missing.

Exploring this a little further, we find that there is little difference in means of obtaining employment in overall sex terms. Seven women and ten men had found their jobs through formal means, six women and seven men had either approached or been approached by the employer, fifteen women and fourteen men had heard of the job from friends or relatives. Turning to the full- or part-time status of the job, only four of the twenty-three part-timers had obtained their job through formal means as compared to thirteen of the forty-two full-timers. Nevertheless the majority of the latter had found employment at the Polytechnic through informal means.

In occupational terms a marked difference appears between the technicians and the other groups. Of the eighteen academic technicians, eleven had heard of their Polytechnic job by newspaper advertisement. One person had transferred from other Council employment, two had been approached by the employer and only four had heard of the job through friends or relatives. This contrasts with the situation in all the other groups. Amongst the maintenance men, two had transferred, one saw the job in a newspaper, and two heard from friends. One of the protection officers obtained his job through the job centre, one via the newspaper, two approached the Polytechnic seeking a job and six heard from friends. One cleaner saw a job advertised, seven made a direct approach, one was asked by the employer, one transferred and eight heard from friends. Two caterers used a newspaper, two transferred and seven heard from friends. One attendant approached the Polytechnic directly and two heard of their job through friends.

Table 3.23 indicates that quite a large proportion of the sample had changed their job since joining their present employer. If we include those who transferred to their present job from other branches of the City Council's employment this applied in twenty-six of the sixty-five cases for whom we have data on this item, (40%).

Table 3.23
Means of obtaining present job.
By sex and occupation.

Sex and occupation	From outside	Internal job transfer	External job transfer	Promotion	Reorgan-isation	Total
Male acad. tech.	6	0	0	1	3	10
Female acad. tech.	4	1	2	1	0	8
Maintenance	3	0	2	0	0	5
Protection	4	0	0	3	3	10
Male cleaners	2	3	0	0	0	5
Female cleaners	11	1	0	1	0	13
Caterers	6	1	0	4	0	11
Attendants	3	0	0	0	0	3
Totals	39	6	4	10	6	65

Note 1: 'From outside' means that the individual was not employed by Coventry C.C. immediately prior to the present job.
Note 2: An 'internal' job transfer is one within the employ of the Polytechnic, an 'external' transfer one from some other Coventry C.C. establishment.
Note 3: In two cases the data was missing.

The table also shows that there is some variation within the sample on this. The women cleaners and attendants have generally moved to their present job from outside the City's employment. In these two cases there has been little internal movement. On the other hand, the protection officers, male cleaners and women technicians have experienced considerable movement whilst in the Council's employ. For the protection officers this is largely to do with the recent reorganisation of the Polytechnic's portering and protection services and the development of an organisational hierarchy within the protection service. For the other two groups, internal movement has been a result of transfer between jobs of similar character. The remaining groups show a degree of internal transfer, the male technicians being involved with reorganisation of the Polytechnic's academic structure, the maintenance men moving from other sections of the Council's direct craft workforce and the caterers gaining promotion, often starting as 'general assistants' in the kitchens or eating places and moving to be cashiers or doing cooking work.

Conclusion.

The sample was constructed with the intention of covering the main ancillary and technical occupations represented at the Polytechnic. These included technical posts in academic departments, maintenance work, the protection service, cleaning, catering and portering/attendant jobs. Alongside this was a concern to include a balance between the sexes and between full and part-time workers. These criteria were used to select the sample. Beyond this, a number of further characteristics of the sample have been established in this chapter. Note that the following overlooks the numerous exceptions to the general picture summarised here. The Polytechnic employs a relatively mature workforce in terms of age. A high proportion of the workers were currently married. Half the sample lived in families with children still at home, these typically being teenagers. Of the married members of the sample a large proportion had spouses in employment, generally in occupations which paralleled their own in terms of their 'white-collar' or 'manual' character.

Sample members had typically joined the Polytechnic in their thirties or forties but employment there was generally stable. People tend to obtain their initial jobs at the Polytechnic by informal means, especially 'hearing of' jobs from friends or relatives already employed there. Once employed, there are opportunities for internal movement between jobs.

Notes.

1. The source here is the **General Household Survey 1983** Table 3.30 Pp. 24 which displays the marital status of persons aged 16 years and over. Note that this includes people over 65 which boosts the proportion widowed and thereby decreases the number married. Nevertheless the point made in the text clearly holds.
2. There is one missing case here.

4 Aggregrate work history

In the Introduction we noted that a prime objective of the research was to examine the past work experience of people currently employed at the Polytechnic. In terms of work, how have they spent their lives since leaving school or college? We turn to this question now, beginning with a discussion of where they began in terms of education and then moving to two forms of analysis of work history. In this chapter we will first consider the sample in *aggregate*, asking what proportion of the sample's *total* time has been spent in various economic activities? This will involve an analysis of time spent in various economic statuses such as paid employment, domestic work or unemployment, and time spent in different industrial sectors, particularly secondary and service industry. Such an aggregate analysis shows us where time has been spent for the sample as a whole but it says little on individuals' pattern of work experience over time, how people move from different statuses, industries and jobs over the course of their working lives prior to coming to the Polytechnic? This will be the concern of chapter five in which we will turn to a scheme of different *patterns of labour market participation over time*, for example, how the work history of men and women differs, or how the work experience of younger workers in the sample differs from older workers? This scheme is derived from the actual patterns of economic activity experienced by the people in the sample. It allows us to take up the question of the extent and nature of *transfer* from the secondary industrial sector to service industry which was the starting point of the research. It also provides the context for the analysis in subsequent chapters of the sample's estimations of their 'best' and 'worst' jobs and how these compare with their present jobs.

Education.

A large proportion of the sample had left school at the minimum school leaving age current at the time. This applied to fifty-nine (88%) of the sample. In terms of the occupational groups, all the maintenance technicians, protection staff, caterers and porters had left school at the minimum age. Two cleaners had remained at school until 17 but the remaining six people were academic technicians, all of whom had remained in education until they were 19 and over.

Again, the majority of the sample had attended elementary, secondary modern or comprehensive school (54 of the 67). This included all the maintenance technicians, caterers and porters. One of the protection men had attended a vocational college and one an 'other' type of school, as had one cleaner. The remaining ten were all academic technicians. Of these, one person had attended a private school, three a grammar school, four had gone to a polytechnic and two to university. These six people who had received higher education prior to entering the labour market were all women technicians whose educational history is thus very different from the rest of the sample.

Turning to qualifications, beginning with those obtained before leaving school or college attended directly after school. Forty-eight of the sixty-seven had left school with no qualifications at all. This applied to all the cleaners, all but two of the caterers (2 had obtained CSE Grades 2-5[1]), all but one of the maintenance men (1 had GCE O level, grades A-C), all but two of the protection men (1 had CSE Grades 2-5, 1 had Scots Lower), and two of the three porters (1 had Scots lower). The academic technicians had rather more qualifications when they left school or college, amongst the men three had CSE Grades 2-5, one had passed O level and one had obtained the School Certificate. This still left five of the ten with no qualifications at the time they left school. By contrast none of the women technicians had left school without qualifications. One had CSE Grade 1, one GCE O level grades A-C, one GCE A level, four had degrees and one had a Diploma in her specialist field which, if taken now, would be of degree standard. So the women technicians stand out in educational terms as compared, not only to the sample as a whole but also to their male colleagues.

We can note at this point that although relatively few of the sample were qualified when they left school, a number of them have obtained qualifications since leaving the education system. Twenty-five people commenced apprenticeships,[2] twenty-one men and four women. The men included nine academic technicians, four maintenance technicians, five protection officers, two cleaners and one attendant. The women included one academic technician and three caterers. Of these twenty-five, seventeen completed an apprenticeship, only one of these was a woman (caterer), the men being eight academic technicians, four maintenance men, three protection officers and one cleaner.

The high incidence of apprenticeship training amongst the male workers is reflected when we turn to qualifications obtained since leaving full-time continuous education. Thirty-two people had obtained or were currently preparing for such a qualification, twenty-three men and nine women. The men included nine of the academic technicians, four maintenance technicians, six protection officers, three cleaners and one of the attendants. The women included three of the academic technicians, four caterers and three cleaners. The qualifications obtained are listed in table 4.1.

A point to note here is the large proportion of 'other' qualifications. This includes a wide variety such as Heavy Goods Vehicle licences, certificates awarded by companies after the completion of a course, marks of competence in tasks within the armed forces and first aid certificates. They tend to have a vocational character as indeed does the range of qualifications in table 4.1, where a large proportion are City and Guilds certificates.

Table 4.1
Post-school/college qualifications obtained.
By sex and occupation.

Sex	Occupation	Qualification	Number
Men	Acad. tech.	GCE 'O' level pass	1
		GCE 'O' level grades D & E	1
		City & Guilds Part 2	2
		City and Guilds part 3	5
		Ordinary National Cert.	1
		Higher National Cert.	1
		Professional	2
		Other	5
	Maintenance	City & Guilds Part 1	2
		City & Guilds Part 2	1
		City & Guilds Part 3	1
		Other	3
	Protection	City & Guilds Part 1	1
		City & Guilds Part 2	1
		Professional	1
		Other	6
	Cleaners	City & Guilds Part 2	1
		Ordinary National Cert.	1
		Other	1
	Attendants	Other	1
Women	Acad. tech.	City & Guilds Part 1	1
		Higher National Cert.	2
		Commercial	1
		Professional	1
		Other	1
	Caterers	City & Guilds Part 1	1
		Commercial	1
		Ongoing	2
	Cleaners	City & Guilds Part 1	1
		Nursing	1
		Other	1

Economic status.

In this section we will provide an analysis of the different economic activities which members of the sample have pursued since they left school or college. The phrase 'economic activities' means the eight different statuses into which any work period was categorised, that is, self-employed, employee, education or training, unemployed, unpaid domestic work, military service, long term sickness and voluntary work.[3] Before turning to this however a methodological point requires note.

As detailed in chapter two, the interview schedule broke down in two cases where the people's work periods were not exhaustive or sequential. As such their work histories were not amenable to the method of analysis undertaken here and they were removed from the data file for present purposes. Again, in one other case, part of the record was missing and this period of that person's work biography has been omitted.

The sixty-five people involved here had between them amassed 1,730 years of 'work' (in the broad sense in which this term is used here) since leaving school or college. This was distributed between the eight statuses as follows:

Table 4.2
Distribution of years by economic status.
Whole sample.

Status	No. of years	% of total
Self employed	28	1.6
Employee	1426	82.4
Educ/training	7	0.4
Unemployed	36	2.1
Domestic work	147	8.5
Military service	75	4.3
Sick	3	0.2
Voluntary work	8	0.5
Total	1730	100.00

Again a methodological note is in order before commenting on table 4.2. It is constructed from the number of years spent in a given work period. In calculating this, rounding was used, so that if a period was less than six months it became the number below and if it was six months or more it became the number above. The consequence of this is that very short spells in a work period became 0 years. This applies to some periods of employment of course but tends to discriminate against some of the categories used which often tend to be of short duration. This applies particularly to sickness and unemployment which may be rather under-represented in table 4.2.

Nevertheless the main import of table 4.2 is quite clear, the vast majority of the sample's working lives has been spent as employees. This is perhaps hardly remarkable, except that certain corollaries follow. Time spent in training, for example, is miniscule as is time spent unemployed when looked at in these aggregate terms. Furthermore, time spent in unpaid domestic work is only 8.5% of the total which is of note given that half the people here are women. To take this a little further, this overall distribution can be broken down by sex as in tables 4.3 and 4.4.

Table 4.3
Distribution of years by economic status.
Men.

Status	No. of years	% of total
Self employed	19	1.8
Employee	940	89.1
Educ/training	2	0.2
Unemployed	17	1.6
Domestic work	0	0.0
Military service	74	7.0
Sick	3	0.3
Voluntary work	0	0.0
Total	1055	100.0

Table 4.4
Distribution of years by economic status.
Women.

Status	No. of years	% of total
Self employed	9	1.3
Employee	486	72.0
Educ/training	5	0.7
Unemployed	19	2.8
Domestic work	147	21.7
Military service	1	0.1
Sick	0	0.0
Voluntary work	8	1.2
Total	675	99.8

The difference in the total number of years between men and women is accounted for by the fact that men in the sample were older than the women. In percentage terms both sex's work histories are dominated by employee status. In the case of men this rises to almost 90%, only military service being at all significant besides this.[4] For women, employee status amounts to 72% of their total working lives. This is lower than the male figure but still represents a predominant life activity. Looked at in these terms being a 'housewife', although it represents over a fifth of the total, comes far behind working for a wage or salary. It is of note that there is no evidence of economic role reversal here, none of the men had spent any time in unpaid domestic labour.

So far I have been looking at the working lives of these people aggregated together. It might be the case that this distorts the picture in that some status has been much more predominant for an individual or sub-group of individuals but this is 'swamped' when totalled in this manner. So let us examine the picture in terms of the proportion of individuals' lives spent in the various statuses.

It is clear from what has been established so far that the majority of people had spent most of their working lives as employees so that it is sensible to use this as a base and then to ask if there are any individuals or subgroups to whom this does not apply? Table 4.5 shows the proportion of *individuals'* adult lives spent in employee status.

Table 4.5
Proportion of individuals' lives spent as employee. Whole sample.

Proportion of time	No. of people	% of people
20 to 29%	1	1.5
30 to 39%	2	3.1
40 to 49%	4	6.2
50 to 59%	6	9.2
60 to 69%	6	9.2
70 to 79%	8	12.3
80 to 89%	12	18.5
90 to 100%	26	40.0
Totals	65	100.0

So 40% of the people had spent over 90% of their working lives as employees, almost 60% over 80% and 70% over 70% of their lives in this way. The mean average proportion of adult life spent as an employee was 79.2%. However, this still leaves a minority who had not spent such large fractions of their lives as employees. Seven people, for example, had spent over 50% of their time outside employee status. We will now go on to briefly examine these other statuses beginning with self-employment.

Five people had spent some of their working life in self-employment, three men and two women. Their businesses had been shopkeeping, plumbing and scrap dealing in the case of the men and shopkeeping and child minding for the women. Apart from the child minder, whose own children had grown beyond the 'minding' stage and who wished to work in adult company after eight years of young children, all the businesses had run into financial difficulties for a variety of reasons. In none of these cases did self-employment represent the bulk of these people's work experience, in percentage terms the largest was a former shopkeeper who had spent 22% of his career in self-employment (10 years).

Turning next to military service, eighteen people had spent some time in the forces, all but one of them men. Five of these had volunteered, the remainder had been conscripted for either war or national service. The conscripts had all spent a relatively short proportion of their total working life in the forces, the one woman had volunteered to make a break with constraining family ties and left after one year when she married. The other volunteers had all spent considerable proportions of their lives in the services. The least amounted to 10% of one man's working life, the others 22%, 39% (both former naval officers) and 89%, a twenty-five year old man who had been in the army from leaving school until 1984.

In the aggregate distribution above, it was mentioned that the method employed discriminated against work periods of less than six months and that this applied to time spent unemployed. This can be corrected now by looking at the unemployment experienced by individuals. Although, by the method used above, only 2.1% of the total years' work of the whole sample was spent unemployed, when this individual method is employed it turns out that 27 of the 65 people had experienced unemployment at some point in their lives. These people are listed in table 4.6 in terms of their age, total number of years unemployed and those years as a percentage of their working lives. It should be remembered that the latter two figures are somewhat nominal as they are the result of the rounding process noted above.

Most people's unemployment has been of relatively short period as a proportion of their total working lives, of the twenty-seven people here, twenty-one had experienced less than 18 month's unemployment through their lives and in eighteen of these cases this amounted to less than 3% of their careers. There are some exceptions to this however. One twenty-one year old had spent 40% of 'working' life unemployed, one of nineteen 33%, one of forty-four 17%, one of twenty-four 25% and one of twenty-eight, 8%.[5] Those with high proportions of unemployment are thus very young compared to the sample as a whole. This is to be expected given the labour market conditions over the last few years. This is emphasised when we turn to quite when these people suffered unemployment. Table 4.7 shows the year in which each period of unemployment began.

Until the late 1970's, the people in the sample experienced very little unemployment. Of the 44 work periods spent in unemployment only 12 occur before 1978. From 1979 however the people in the sample have suffered 32 periods of unemployment.

We can now turn to the incidence of unemployment experience in terms of the current occupation of the people concerned. This is listed in table 4.8. The maintenance technicians had clearly the least experience of unemployment, none of the five had been unemployed. Only two of the male academic technicians had been unemployed and their unemployment had been for a small proportion of their lives. In terms of the proportion of the occupational subgroup who had been unemployed at some point, the protection men, porters and male cleaners all show a high number with unemployment experience (7 out of 10, 2 out of 3 and 4 out of 5 respectively). In each case, however, their unemployment represented a fairly small fraction of their total working lives. In the case of women, four of the thirteen cleaners and four of the eleven caterers had been unemployed at some point although there is a marked difference between the proportions of their lives spent unemployed. The difference is exaggerated by the single case of the woman cleaner who defined her years looking after her young children as 'unemployment' but the difference remains. The women academic technicians stand out. Four of the eight people had unemployment experience and it averaged 16.7% of their working lives. Unlike the men, where qualifications and skills protected the individual from unemployment, these highly qualified women had the severest experience of unemployment in the sample.

Table 4.6
Those with unemployment experience.
Age (1985), years unemployed and unemployment as a percent of working life.

Age	Years	% of life
19	1	33.3
21	2	40.0
24	0	0
24	1	25
25	0	0
25	0	0
28	1	8.3
31	1	6.3
32	0	0
33	0	0
36	12	60
39	0	0
44	5	16.7
49	1	2.9
49	1	2.9
51	2	5.6
52	1	2.7
53	0	0
55	2	4.9
57	0	0
59	2	4.4
60	0	0
60	1	2.2
61	1	2.3
62	1	2.2
63	1	2.1
64	0	0

Table 4.7
Year in which a work
period of unemployment began.

Year	No. of work periods
1958	1
1965	1
1966	1
1967	1
1970	2
1971	1
1972	1
1973	1
1974	1
1977	2
1979	4
1980	7
1981	5
1982	7
1983	6
1984	1
1985	2

Table 4.8
Unemployment experience (% of adult years
unemployed) by sex and occupation.

Sex and occupational group	No. in sample	No. with unemployment experience	% of adult years unemployed
Male acad. tech.	10	2	0.0
Female acad. tech.	8	4	16.7
Male maint. tech.	5	0	0.0
Male protection	10	7	2.3
Male cleaners	5	4	2.8
Female cleaners	13	4	26.3
Female caterers	11	4	0.0
Male porters	3	2	11.1
Total	65	27	8.2

It has already been noted that only the women workers in the sample had experience of unpaid domestic work. Twenty-one of the thirty-two women included here had spent some part of their lives in this capacity, that is, eleven of the women had never been full-time 'housewives'. Of those who had, the average percent of their working lives spent this way was 26.3. This varied from less than one percent to 50%. As it happens, the one woman who had spent 50% of her working life looking after her home and family was very untypical as she was the only woman technician who had been a full-time housewife. On the other hand, most of the women cleaners (11 out of 13) and caterers (9 out of 11) had spent some time at home. The caterers had averaged 27.8% of their adult lives as housewives, the cleaners 22.9%. This is not, however, a significant difference when their respective ages are noted, the cleaners being somewhat older than the caterers.

It has already been noted that, in aggregate, other economic statuses are very minor. In terms of individual experience, eleven people had spent some time in education or training, long-term sickness or voluntary work. For seven of these, such periods had been minor parts of their total working lives (7% and under). Four people, however, had spent significant proportions of their adult lives in one of these activities. A twenty-two year old porter had spent 14% of his time since school on a training scheme, a twenty-one year old cleaner had spent 20% of her life on such a scheme. One woman technician of twenty-eight had returned to education to improve her qualifications spending 25% of her adult life in education. Finally, one woman caterer had worked as a voluntary helper in her children's school, this amounting to 30% of her working life.

Industrial background.

We now turn to the industries in which these people have worked during the course of their adult lives. Of course the range of products and services produced in economic activity is enormous and some classificatory scheme is necessary to order the information. The basic scheme used in this project is the Labour Force Survey's list of industries. However, this is still too unwieldy for present purposes as it contains approximately 400 different industries. Hence the industrial categories have been amalgamated into fourteen industrial sectors to begin the analysis. This scheme is a modified version of the Central Statistical Office's *Standard Industrial Classification 1980* (S.I.C.)[6] The categories employed are:

1.**Agriculture.** S.I.C. 0. Agriculture, forestry and fishing. (Labour Force Survey industrial codes 500 to 502).

2.**Utilities** S.I.C. 1. Energy and water supply industries. (L.F.S. codes 511 to 513).[7]

3.**Manufacturing.** S.I.C.'s 3. Metal goods, engineering and vehicle industries and 4. Other manufacturing industries. (L.F.S. codes 516 to 522, 526 to 724). The two S.I.C. manufacturing divisions are amalgamated here as when more detailed attention is required we will move to the particular industries concerned.

4.**Construction.** S.I.C. 5. Construction. (L.F.S. code 725). Note that local government employees are included here insofar as they are involved in construction work.

5.**Distribution.** S.I.C. 6. Distribution, hotels and catering, repairs. (L.F.S. codes 726 to 741, 771)

6.**Transport.** S.I.C. 7. Transport and communication. (L.F.S. codes 742 to 754).

7.**Commerce.** S.I.C. 8. Banking, finance, insurance, business services and leasing. (L.F.S. codes 755 to 770, 772 to 770, 790).

8.**Other services.** This comprises S.I.C. division 9 Other Services with the exception of numbers 9 and 10 below. It employs L.F.S. codes 776 to 779, 781 to 784, 786 to 789, 791 to 810. This rather enigmatically titled group encompasses largely the public sector services such as those provided by the health authorities and local and central government. However, it is rather broader than this in including such services which are provided within the private sector, for example, fee paying schools, as well as a range of miscellaneous services such as L.F.S. code 799 'Religious organisations and similar associations' and 809 'Personal services not elsewhere classified, e.g. funeral services, piano tuners, kennels'!

9.**Polytechnics.** (L.F.S. code 785) This industry is picked out for special attention as all the people in the survey are currently employed in it.

10.**Military service.** (L.F.S. code 780) This industry is picked out for special attention as many of the people included in the sample are of an age whereby service in the armed forces has formed part of their work history due to conscription for wartime or national service.

As well as these conventional industrial categories the following analysis also includes a number of unwaged 'economic' activities, namely:

11.**Unpaid domestic labour.**

12.**Unemployment.**

13.**Other non-paid activities.** I.e. education or training, long term sickness and voluntary work.

14.**Casual work.** This covers periods of time when an individual worked casually in a variety of industries without having a permanent job of his or her 'own'.

We can again look at the overall picture of where the sample had spent their working lives, this time in terms of these industrial sectors.

Table 4.9
Distribution of years by industrial sector.
Whole sample.

Industrial sector	No. of years	% of total	Total in service industry
Agriculture	9	0.5	
Utilities	3	0.2	
Manufacturing	597	34.5	
Construction	106	6.1	
Distribution	148	8.6	)
Transport	11	0.6	)
Commerce	12	0.7	)41.6
Other services	144	8.3	)
Polytechnic	405	23.4	)
Military service	75	4.3	
Domestic work	147	8.5	
Unemployed	36	2.1	
Other non-paid	17	1.0	
Casual work	20	1.2	
Total	1730	100.0	

The first point to note here is the proportion of time spent in the secondary sector, around 40% of the total. One of the principal aims of the project was to discover whether there had been a transfer of people from the secondary to the service sector. The fact that two-fifths of the sample's time had been spent in manufacturing and construction suggests that there has indeed been such a transfer although at the present stage of analysis this is only suggestive and we will return to it later. Second, putting the service industries together (as in the right hand column of table 4.9) they amount in total to two-fifths of the sample's working lives. But it is also noticeable here that a large proportion of this time in service industry has been spent at the Polytechnic itself.[8]

Still remaining with the aggregate years, we can go on to examine how these vary in terms of sex and occupational group. The distribution of years for men and women is shown in tables 4.10 and 4.11. In the case of men, the proportion of total time spent in the secondary sector rises to 45% and the time spent in service industry falls to just over a third. It is also noticeable that the Polytechnic is by far the most important service industry, men's experience of other service sector work is very limited indeed. For women, experience of secondary industry is much less than men, falling to under a fifth of the total, whilst time in service industry rises to over a half. Furthermore, although the Polytechnic accounts for a sizable proportion of service sector work, this is not so predominant as in the men's case, in that women have much larger proportions of time spent in distribution and other services. It can also be noted here that none of the women's time had been spent in construction whilst this amounted to 10% of the men's.

Moving now to the occupational groups, displayed in tables 4.12 to 4.17. The academic technicians have the average manufacturing experience of the whole sample but apart from this their work time has been dominated by the Polytechnic which amounts to over two fifth's of the total. Time spent in other service sector industries is minimal. The maintenance technicians' past work has been predominantly in the construction sector with relatively little manufacturing and service experience. Furthermore employment in the service sector has been almost totally at the Polytechnic itself. The protection men have a relatively high rate of manufacturing experience and a low rate of service work. Their service employment is much less concentrated at the Polytechnic than any of the other groups. In general their industrial experience is more spread across the various sectors. Bearing in mind that many of the cleaners are women their experience of manufacturing industry is quite extensive. Service employment is about the average but Polytechnic employment is not so predominant here as in other cases. The caterers have a very low rate of manufacturing experience and the highest rate of service sector employment. A relatively large proportion of their employment has been spent at the Polytechnic. There are of course only three attendants involved here but clearly their work has been predominantly in the manufacturing sector with very little service experience and that exclusively at the Polytechnic.

Table 4.10
Distribution of years by industrial sector.
Men.

Industrial sector	No. of years	% of total	Total in service industry
Agriculture	9	0.8	
Utilities	2	0.2	
Manufacturing	478	45.3	
Construction	106	10.0	
Distribution	39	3.7	)
Transport	8	0.8	)
Commerce	2	0.2	)34.5
Other services	51	4.8	)
Polytechnic	264	25.0	)
Military service	74	7.0	
Domestic work	0	0.0	
Unemployed	17	1.6	
Other non-paid	4	0.4	
Casual work	1	0.1	
Total	1055	99.9	

Table 4.11
Distribution of years by industrial sector.
Women.

Industrial sector	No. of years	% of total	Total in service industry
Agriculture	0	0.0	
Utilities	1	0.1	
Manufacturing	119	17.6	
Construction	0	0.0	
Distribution	109	16.1	)
Transport	3	0.4	)
Commerce	10	1.5	)52.7
Other services	93	13.8	)
Polytechnic	141	20.9	)
Military service	1	0.1	
Domestic work	147	21.8	
Unemployed	19	2.8	
Other non-paid	13	1.9	
Casual work	19	2.8	
Total	675	99.9	

Table 4.12
Distribution of years by industrial sector.
Academic technicians.

Industrial sector	No. of years	% of total	Total in service industry
Agriculture	0	0.0	
Utilities	3	0.8	
Manufacturing	134	34.6	
Construction	22	5.7	
Distribution	2	0.5	)
Transport	0	0.0	)
Commerce	0	0.0	)48.1
Other services	18	4.7	)
Polytechnic	166	42.9	)
Military service	23	5.9	
Domestic work	12	3.1	
Unemployed	3	0.8	
Other non-paid	3	0.8	
Casual work	1	0.3	
Total	387	100.1	

Table 4.13
Distribution of years by industrial sector.
Maintenance technicians.

Industrial sector	No. of years	% of total	Total in service industry
Agriculture	0	0.0	
Utilities	0	0.0	
Manufacturing	18	13.4	
Construction	74	55.2	
Distribution	1	0.7	)
Transport	0	0.0	)
Commerce	0	0.0	)27.5
Other services	3	2.2	)
Polytechnic	33	24.6	)
Military service	5	3.7	
Domestic work	0	0.0	
Unemployed	0	0.0	
Other non-paid	0	0.0	
Casual work	0	0.0	
Total	134	99.8	

Table 4.14
Distribution of years by industrial sector.
Protection staff.

Industrial sector	No. of years	% of total	Total in service industry
Agriculture	7	2.2	
Utilities	0	0.0	
Manufacturing	155	48.1	
Construction	9	2.1	
Distribution	24	7.4	)
Transport	8	2.5	)
Commerce	2	0.6	)33.7
Other services	32	9.9	)
Polytechnic	43	13.3	)
Military service	35	10.9	
Domestic work	0	0.0	
Unemployed	5	1.5	
Other non-paid	2	0.6	
Casual work	0	0.0	
Total	322	99.8	

Table 4.15
Distribution of years by industrial sector.
Cleaning staff.

Industrial sector	No. of years	% of total	Total in service industry
Agriculture	2	0.3	
Utilities	0	0.0	
Manufacturing	209	35.8	
Construction	0	0.0	
Distribution	84	14.4	)
Transport	0	0.0	)
Commerce	0	0.0	)41.9
Other services	68	11.6	)
Polytechnic	93	15.9	)
Military service	12	2.0	
Domestic work	75	12.8	
Unemployed	21	3.6	
Other non-paid	1	0.2	
Casual work	19	3.2	
Total	584	99.8	

Table 4.16
Distribution of years by industrial sector.
Catering staff.

Industrial sector	No. of years	% of total	Total in service industry
Agriculture	0	0.0	
Utilities	0	0.0	
Manufacturing	28	12.1	
Construction	0	0.0	
Distribution	37	16.0	)
Transport	3	1.3	)
Commerce	10	4.3	)58.0
Other services	23	10.0	)
Polytechnic	61	26.4	)
Military service	0	0.0	
Domestic work	60	26.0	
Unemployed	0	0.0	
Other non-paid	9	3.9	
Casual work	0	0.0	
Total	231	100.0	

Table 4.17
Distribution of years by industrial sector.
Attendants.

Industrial sector	No. of years	% of total	Total in service industry
Agriculture	0	0.0	
Utilities	0	0.0	
Manufacturing	53	73.6	
Construction	1	1.4	
Distribution	0	0.0	)
Transport	0	0.0	)
Commerce	0	0.0	)12.5
Other services	0	0.0	)
Polytechnic	9	12.5	)
Military service	0	0.0	
Domestic work	0	0.0	
Unemployed	7	9.7	
Other non-paid	2	2,7	
Casual work	0	0.0	
Total	72	99.9	

So far the analysis of industrial background has been cast in terms of industrial sectors. We can now briefly move to the particular industries in which the people in the sample have worked. Table 4.18 gives the full list for the sake of completeness along with each industry's Labour Force Survey code number, the number of work periods spent in an industry and its percentage of the total number of work periods.

There are ninety-eight different industries here. However, although this indicates the range of industrial experience of the people in the sample, it does not accurately measure the relative importance of the industries listed. This is because it represents a count of *work periods* which can vary in length from a matter of days to a number of years. However, table 4.18 can be used as a basis for more detailed analysis in suggesting those industries which are *prima facie* important in that they contain a number of work periods. If we take five to be the criterion here, this gives us the industries listed in table 4.19. This table also shows the number of years spent in each industry and as this adds up to 1245, that is, 73% of the sample's total number of years, we can be confident that all the important industries in these people's work history have been included.

A number of points can be made from table 4.19. As the Polytechnic is the current employer of the people in the sample, one would expect it to have a degree of prominence. However, it is worth repeating that almost a quarter of the sample's total working years have been spent at the Polytechnic. This confirms the point made in chapter three regarding the length of service with the present employer. Second, Coventry's major manufacturing industries are well represented here. Putting together industries connected to vehicle manufacture, aerospace, mechanical and electrical industries they amount to 408 working years. Most prominent here is, of course, vehicle manufacture itself. A point to note, which will be developed in chapter five, is the relatively large number of years per work period for time spent in vehicle manufacture (9.2 compared to the average of 3.4). Whilst mentioning leading industries one which is marked by its absence is textiles, notably the man-made fibre industry. Although this declined severely in the 1970's (not an uncommon occurrence in Coventry's manufacturing!) it still employed 5,090 people in 1966 and 3,760 in 1971.[9] Yet none of the people in the sample appeared to have worked in it. We can offer no explanation for this, only speculate that as an industry in long term decline it employed a relatively old workforce which has since withdrawn from the labour market.

The pattern of participation in these leading industries in terms of sex is shown bin table 4.20. The first point to note here is the disparity in the totals. Of the 1261 years worked in these most prominent industries, 914, that is 72%, have been worked by men. As the measure of importance used here is the number of years worked, this suggests that men have worked in more stable jobs than women, staying in a job for a longer period of time. This is confirmed by table 4.21 which shows the average number of years spent in a job in these twenty-one most prominent industries. The average length of time spent by women in these industries is thus less than half that spent by men. This characteristic is quite consistent across the range. In only two industries, hospitals and vehicle body manufacture is the relationship reversed. This feature is clearly linked to the kind of industries in which men and women have worked because, as can be seen from table 4.22, there is a high degree of segregation in the industries where men and women have spent their working lives.[10]

Table 4.18
Industries in which at least one work period has been spent. Labour Force Survey code, number of work periods and proportion of total work periods.

Industry	L.F.S. code number	No. of w.p.'s	Percent of total w.p.'s
AGRICULTURE	500	3	.6
ELECTRICITY PRODUCTION	511	1	.2
WATER SUPPLY	514	1	.2
STEEL PRODUCTION	516	3	.6
ALUMINIUM PRODUCTION	520	1	.2
COPPER PRODUCTION	521	2	.4
CEMENT ETC PRODUCTION	527	1	.2
FLAT GLASS MANUFACTURE	533	1	.2
GLASS CONTAINER MANUFACTURE	534	1	.2
PHARMACEUTICAL PRODUCTION	553	1	.2
STAINLESS STEEL PRODUCTS	566	2	.4
METAL FURNITURE MANUFACTURE	570	1	.2
OTHER METAL PRODUCTS	572	2	.4
AGRICULTURAL MACHINE MANUFACTURE	576	7	1.3
MACHINE TOOL MANUFACTURE	577	5	.9
CONSTRUCTION EQUIPMENT M/URE	584	3	.6
OTHER MECHANICAL ENGINEERING	597	2	.4
ELECT ENG, POWER EQUIPMENT	602	5	.9
ELECT ENG, VEHICLE PARTS	605	1	.2
ELECT ENG, TELEPHONES	607	14	2.6
ELECT ENG, COMPONENTS	610	2	.4
ELEC ENG, CONSUMER GOODS	613	5	.9
ELEC ENG, DOMESTIC APPLIANCES	614	1	.2
ELECT ENG, LIGHTING	615	1	.2
ELECTRICAL INSTALLATION	616	5	.9
VEHICLE MANUFACTURE	617	26	4.9
VEHICLE BODY MANUFACTURE	618	5	.9
VEHICLE COMPONENTS METAL	621	10	1.9
SHIPBUILDING	622	1	.2
RAILWAY STOCK MANUFACTURE	623	3	.6
MOTOR CYCLE MANUFACTURE	624	1	.2
AEROSPACE	626	7	1.3
MEDICAL EQUIPMENT	629	2	.4
SPECTACLES-LENS MANUFACTURE	630	2	.4
FRUIT-VEGETABLE PROCESSING	641	1	.2
BREAD-CONFECTIONERY PRODUCTION	645	1	.2
CARPET MANUFACTURE	667	1	.2
RIBBON MANUFACTURE	671	2	.4
LEATHER GOODS MANUFACTURE	674	1	.2
MEN'S TAILORED CLOTHING	677	1	.2
MEN'S SHIRTS ETC	680	1	.2
WOMEN'S NONTAILORED CLOTHING	681	3	.6
OTHER CLOTHING MANUFACTURE	684	1	.2
WOODEN FURNITURE MANUFACTURE	696	1	.2
NEWSPAPER PRODUCTION	705	1	.2
OTHER PRINTING-PUBLISHING	708	8	1.5
TYRE MANUFACTURE	709	1	.2

PLASTIC SEMI MANUFACTURED GOODS	713	2	.4
OTHER PLASTIC GOODS	717	1	.2
JEWELLERY MANUFACTURE	718	2	.4
PHOTOGRAPHIC PROCESSING	720	1	.2
TOY & GAME MANUFACTURE	721	3	.6
OTHER MANUFACTURE	724	1	.2
CONSTRUCTION	725	30	5.6
WHOLESALE DISTRIBUTION-CARS ETC	726	1	.2
WHOLESALE DISTRIBUTION	727	4	.8
SCRAP DEALING	728	1	.2
RETAIL DISTRIBUTION-CARS ETC	730	3	.6
RETAIL DISTRIBUTION	731	33	6.2
CAFE-RESTAURANTS	732	9	1.7
TAKE AWAY FOOD	733	1	.2
PUBS & BARS	734	1	.2
LICENSED CLUBS	735	2	.4
CANTEENS INCL. SCHOOLS	736	3	.6
HOTELS	737	6	1.1
CAR REPAIRERS	739	1	.2
ROAD HAULAGE	745	2	.4
SEA TRANSPORT	747	1	.2
CAR PARKS	749	1	.2
POSTAL SERVICES	753	1	.2
TELECOMMUNICATIONS	754	2	.4
BANKING	755	1	.2
SOLICITORS	761	1	.2
OTHER PROFESSIONAL SERVICES	763	1	.2
OTHER BUSINESS SERVICES	766	4	.8
OTHER LOCAL GOVERNMENT	776	6	1.1
FIRE SERVICE	779	1	.2
NATIONAL DEFENSE	780	17	3.2
DHSS	781	1	.2
REFUSE DISPOSAL	782	1	.2
SEWAGE DISPOSAL	783	1	.2
CLEANING SERVICES	784	2	.4
POLYTECHNICS	785	102	19.1
SCHOOLS	786	8	1.5
TECHNICAL COLLEGES	788	3	.6
HOSPITALS-NURSING HOMES	791	10	1.9
LOCAL GOVT WELFARE SERVICES	797	3	.6
FILM PRODUCTION-EXHIBITION	801	2	.4
ENTERTAINMENTS	802	2	.4
DRY CLEANING	807	2	.4
HAIRDRESSING	808	1.2	
OTHER PERSONAL SERVICE	809	1	.2
DOMESTIC SERVICE	810	7	1.3
NON-PAID WORK PERIODS		93	17.5
TOTAL		533	100.0

Table 4.19
The 21 most prominent industries.

Industry	Total no. years	No. of work periods	Years per work period
Polytechnic	405	102	4.0
Vehicle manufacture	240	26	9.2
Construction	106	30	3.5
Retail distribution	84	33	2.5
National defense	70	17	4.1
Aerospace	45	7	6.4
Hospitals	39	10	3.9
Other printing and publishing	39	8	4.9
Elect eng, telephones	36	14	2.6
Vehicle components	30	10	3.0
Schools	25	8	3.1
Other local government	20	6	3.3
Agricultural machine m/ture	19	7	2.7
Elect eng, power machinery	18	5	3.6
Electrical installation	16	5	3.2
Vehicle body manufacture	16	5	3.2
Domestic service	15	7	2.1
Machine tool manufacture	13	5	2.6
Cafe-restaurants	9	9	1.0
Hotels	9	6	1.5
Elect. eng, consumer goods	7	5	1.4
Total	1261		
Average			3.4

Table 4.20
The 21 most prominent industries.
Total number of years by sex.

Industry	Men no. of years	Women no. of years.
Polytechnic	264	141
Vehicle manufacture	233	7
Construction	106	0
Retail distribution	25	59
National defense	69	1
Aerospace	42	3
Hospitals	19	20
Other printing and publishing	35	4
Elect eng, telephones	14	22
Vehicle components	27	3
Schools	14	11
Other local government	0	20
Agricultural machine m/ture	17	2
Elect eng, power machinery	18	0
Electrical installation	16	0
Vehicle body manufacture	4	12
Domestic service	0	15
Machine tool manufacture	11	2
Cafe-restaurants	0	9
Hotels	0	9
Elect. eng, consumer goods	0	7
Total	914	347

Table 4.21
The 21 most prominent industries.
Average number of years per work period by sex.

Industry	Men average years	Women average years.
Polytechnic	4.6	3.2
Vehicle manufacture	9.7	3.5
Construction	3.5	
Retail distribution	3.1	2.4
National defense	4.3	1.0
Aerospace	8.4	1.5
Hospitals	3.8	4.0
Other printing and publishing	7.0	1.3
Elect eng, telephones	7.0	1.8
Vehicle components	3.4	1.5
Schools	7.0	1.8
Other local government		3.3
Agricultural machine m/ture	5.7	2.0
Elect eng, power machinery	3.6	
Electrical installation	3.2	
Vehicle body manufacture	1.3	6.0
Domestic service		2.5
Machine tool manufacture	2.7	1.0
Cafe-restaurants		1.0
Hotels		2.2
Elect. eng, consumer goods		1.4
Total	4.6	2.2

Table 4.22
The 21 most prominent industries.
Percent of years in an industry, by sex.

Industry	Men % years	Women % years.
Polytechnic	65.2	34.8
Vehicle manufacture	97.1	2.9
Construction	100.0	0.0
Retail distribution	29.8	70.2
National defense	98.6	1.4
Aerospace	93.3	6.7
Hospitals	48.7	51.3
Other printing and publishing	89.7	10.3
Elect eng, telephones	38.9	61.1
Vehicle components	90.0	10.0
Schools	56.0	44.0
Other local government	0.0	100.0
Agricultural machine m/ture	89.5	10.5
Elect eng, power machinery	100.0	0.0
Electrical installation	100.0	0.0
Vehicle body manufacture	25.0	75.0
Domestic service	0.0	100.0
Machine tool manufacture	84.6	15.4
Cafe-restaurants	0.0	100.0
Hotels	0.0	100.0
Elect. eng, consumer goods	0.0	100.0

Table 4.22 attempts to make this point clear by converting the numbers of table 4.20 into the percentage of years spent in an industry as between men and women. Thus a lack of segregation by sex is indicated by percentages which are roughly equal, whilst the wider apart the two figures the greater the degree of segregation. We can split the industries into the following subgroups.

Industries with a high degree of segregation, large proportion of men and small proportion of women:

Vehicle construction.
Construction.
National defense.
Aerospace.
Other printing / publishing.
Vehicle components.
Agricultural machine manufacture.
Electrical engineering, power machinery.
Electrical installation.
Machine tool manufacture.

Clearly, with the exception of the military services, all of these industries in which men have predominantly worked lie in the secondary sector.

Industries with a high degree of segregation, large proportion of women and small proportion of men:

Domestic service.
Cafe-restaurant.
Hotels.
Other local government.
Electrical engineering, consumer goods.

By contrast most of these industries lie in the service sector. However in some other industries there is not such a clear cut division between men in service industry and women in services. In the Polytechnic, in the service sector, there is a predominent of men but with a substantial minority of women. In retail distribution, electrical engineering (telephone manufacture) and vehicle body manufacture, a mix of services and manufacturing, there is a majority of women but with a minority of men. In only two industries, hospitals and schools, both public sector services on the whole, is there anything like an absence of segregation by gender. We are of course here looking simply at the industries people have worked in, not the jobs they have done within them.

We now come to the question of when people worked in the industrial sectors we are using here, again looked at in this aggregate fashion. We will do this by outlining the work periods that fall into the six decades from the 1930's to the 1980's. Of course, taking decades is somewhat arbitrary but the intention is to see whether in general terms the people in the sample have shifted over from one sector to another and if so when this occurred. Again work periods do not fall neatly into the decades so that there is considerable duplication here. For example, a work period beginning in 1958 and ending in 1963 will appear in the discussion of both the 1950's and the 1960's. To repeat, the intention is to assess the overall pattern of work done by the people in the sample during this period of time. More detailed examination of the process of change in individuals' work biographies will follow later in chapter five.

Table 4.23
Percent of work periods in industrial sectors in 6 decades.

Sector	1930's	1940's	1950's	1960's	1970's	1980's
Agriculture	11.1	1.6	1.1	0.6	0.0	0.0
Utilities	0.0	0.0	0.0	0.6	0.0	0.5
Manufacturing	44.4	46.0	49.5	40.5	26.3	12.4
Construction	0.0	7.9	10.5	9.5	3.8	1.6
Distribution	33.3	15.9	9.5	13.1	15.1	5.9
Transport	0.0	1.6	2.1	2.4	0.5	0.5
Commerce	0.0	1.6	0.0	0.6	1.6	2.2
Other services	0.0	4.8	9.5	11.3	11.8	8.6
Polytechnic	0.0	0.0	1.1	5.4	19.9	45.2
All services	33.3	23.9	21.2	32.8	47.5	62.4
Military service	11.1	19.0	8.4	1.2	0.5	0.5
Domestic work	0.0	1.6	6.3	11.9	10.2	2.2
Unemployed	0.0	0.0	1.1	1.8	7.0	16.7
Other non-paid	0.0	0.0	1.1	0.6	2.2	3.2
Casual work	0.0	0.0	0.0	0.6	1.1	0.5
Total	100.0	100.0	100.0	100.0	100.0	100.0

Table 4.23 can only be regarded as indicative as a number of considerations slant the figures. In the case of the 1930's decade, only a few work periods are included, the 1940's is clearly influenced by war service and for the 1980's the fact that all the people in the sample worked in the Polytechnic obviously reduces everything else. However, a number of points are evident. First, manufacturing and the total of services together make up the majority of the work periods covered. If we add them together they consistently amount to between 70 and 78% of the total throughout the period covered.[11] However, the relative weight of manufacturing and services clearly changes over time. Leaving the 1930's aside manufacturing amounts to twice the number of work periods as services in the 1940's and 1950's, then manufacturing declines steadily in the 1960's and 1970's with a corresponding rise in the services. The increase in the latter is not, however, spread evenly across the different service industries. The numbers involved in transport and commerce have consistently been very small, the proportion working in distribution is larger, but there is no consistent pattern of growth here. The expansion in the services has clearly occurred in the 'other services' sector, if we for the moment include the Polytechnic in that as it usually is. The figures here then become 4.8% of work periods in the 1940's, 10.6% in the 1950's, 16.7% in the 1960's and 31.7% in the 1970's. The main shift, in these aggregate terms, has thus been from manufacturing to the public services, a shift which, again just from the point of view of these aggregate figures, begins around 1960 and has gone on since.

Conclusion.

This chapter began with a brief discussion of the sample's education. Most of the people left school at the current minimum leaving age with very few qualifications. A number of them have obtained qualifications since leaving school, generally vocationally oriented and often linked to apprenticeship. The women technicians are exceptional in terms of the type of educational institution attended, the age they left education and their level of qualifications.

Turning to the economic positions which the sample members had held over their working lives, the great majority of their time had been spent as employees. By comparison, the other economic statuses were of minor significance in these *overall* terms. However, in terms of the proportion of *individuals'* lives spent in such statuses, rather more significance is suggested. For example, although being a housewife forms a minor part of women's total working life, most women had spent some time doing this. At this stage, though, we cannot say *when* this occurs and *how* it fits into the sequence of working life. Again, although in overall terms, time spent unemployed was rather small, a large minority of the sample had been unemployed at some stage. Such points require the analysis of the pattern of participation of individuals over time which we turn to in the next chapter.

Moving to the industrial background of the sample. In terms of industrial sectors, a large proportion of the people's total time in work has been spent in the secondary sector and at the Polytechnic. The former amounted to 41% of the total, the latter 24%. These statistics indicate the possibility of transfer from the secondary sector to the Polytechnic but we also noted a series of variations between the sexes and occupations in these proportions of overall time in industrial sectors. Once again we need to proceed to examine when people worked where and how their participation in different industrial contexts fits into the pattern of their working life.

In terms of particular industries, the sample members had worked in ninety-eight different industries, although twenty-one stand out as most prominent, particularly Coventry's major manufacturing areas. Analysis of these showed considerable variation, particularly by sex, we we shall pursue in chapter five.

The chapter concluded with an analysis of when the sample as a whole had worked in different industrial sectors. This again suggests a move during the course of their working lives from manufacturing/construction to public service industry. We can close this chapter with reference to tables 4.24 and 4.25 which reinforce this. These show that thirty-one[12] of the thirty-four men included here (i.e. 88%) and twenty-five of the thirty-three women (i.e. 76%) had, at some time, worked in the secondary sector. As well as such experience being spread between the sexes it is also apparent that most of the occupational groups include a majority of people who have worked in manufacturing and/or construction. However, such simple figures do not indicate to us the nature of that experience, for example, how long it is, whether secondary sector and service employment are interspersed through life or whether there has been a transfer from one sector to the other? To answer these kinds of questions we move to an analysis of individuals' patterns of work history.

Table 4.24
Male workers who have worked in secondary industry, by current occupational group.

Occupational Group	Those with secondary experience	Those without secondary experience
Academic tech	8	2
Mainten tech	5	0
Protection	9	1
Cleaners	6	0
Attendants	3	0
Total	31	3

Table 4.25
Female workers who have worked in secondary industry, by current occupational group.

Occupational Group	Those with secondary experience	Those without secondary experience
Academic tech	6	2
Cleaners	12	1
Caterers	7	5
Total	25	8

Notes.

1. The questionnaire did not ask for details of all qualifications obtained but only the highest level of attainment. This is what is indicated in this paragraph.
2. The discussion of quite what constitutes an apprenticeship in chapter two should be borne in mind here.
3. See chapter two.
4. Many of the men in the sample had been subject to conscription either in war time or for National Service.
5. I have omitted the case of the person who had spent 60% of her life unemployed as this was a woman who defined the twelve years spent whilst her children were young as being unemployed. Her self definition of this period in these terms is of interest but rather incompatible with the rest, so far as is known she was not looking for employment nor willing to accept employment in this period.
6. Central Statistical Office **Standard Industrial Classification 1980** H.M.S.O. 1979.

7. Note that the S.I.C. division 2, extractive industries is omitted for the simple reason that none of the sample had worked in any of the industries categorised there.

8. The category 'Polytechnic' does not include only the Coventry (Lanchester) Polytechnic but Polytechnics in general but the fact is that almost without exception it is the Coventry Polytechnic which is referred to here.

9. Census figures taken from I. Procter (1984a: 76)

10. Compare with S. Tolliday, (1986: 206-207).

11. The actual totals are: 1930's (77.7%), 1940's (69.9%), 1950's (70.7%), 1960's (73.3%), 1970's (73.8%), 1980's (74.8%).

12. This figure is inconsistent with the claim forwarded in the next chapter that four, rather than three, men had worked all their lives in the service sector. The disparity is accounted for by one man who worked very briefly in manufacturing before World War II but who has since worked in services. In the next chapter his few months in manufacturing are ignored.

5 Patterns of labour market participation

Introduction.

Clearly each individual's biography of 'work' is *individual* to them just as each person's experience of health and illness is specific to themselves. Yet just as we can detect wider patterns in the incidence of health and disease over people's lives (the tendency of women to live longer than men is the simplest example) so we can also uncover a patterning in the course of people's working lives. In this introductory section we will outline what seem to be the main determining factors in shaping the work biographies of the members of the sample. At this stage these will simply be stated to indicate the overall shape of the detailed analysis which follows.

The first factor to introduce is age. We noted in chapter three that the Polytechnic's workforce tends to be relatively old and that employment at the Polytechnic tends to commence in people's thirties or forties. Yet there is a significant minority of younger workers and we will establish that these workers have pattern a of labour market experience which differs from that of their older colleagues when they were young workers. The present cohort of workers under thirty years of age have a biography of work marked by a) the instability of employment prior to entering the present job, b) a process of gradually 'edging in' to a more secure position in the labour market and c) very limited experience of manufacturing employment. As such we will separate them off into a group of *younger* workers.

The second criterion to introduce is *sex* on the grounds that men and womens' labour market experience differed systematically in four respects:

a) A large proportion of the older women had interrupted their participation in the paid employment system for a 'domestic interlude' usually associated with the care of young children. None of the men had done this.

b) This break was an 'interlude', varying in length and singularity, but on return to employment many women had obtained part-time employment. Again none of the men had held part-time jobs.

c) As we have already established, the industries in which men and women had worked show a marked pattern of segregation, men working in some, women in others.

d) This was not so simple as men in secondary industry, women in services. We have noted that 75% of women had worked in secondary industry. But the pattern of participation is different. The older men in the study had spent the bulk of their working lives in manufacturing/construction prior to coming to the Polytechnic. The jobs they had held there were much longer term than the womens' who had generally worked in secondary industries prior to their domestic interlude.

So for *women* the dominant pattern of labour market experience was 1) full-time employment between school/college and a domestic interlude. This includes a mix of jobs in services and the secondary sector. 2) A domestic interlude. 3) Part-time employment, largely in services. There is thus a definite shift in the type of jobs women hold, largely in connection with their continuing domestic responsibilities when they return to paid employment after an interlude. We will call this group of workers *post-domestic women*, to sum up their labour market experience although the use of this term does not at all imply that the domestic responsibilities of the women involved have ceased. It is also of note that a small number of older women had not followed this pattern as they had not taken a domestic break from paid employment. We will keep these people separate and refer to them as the *non-domestic* women.

In the case of the majority of *men* the great bulk of their employment prior to coming to the Polytechnic had been in manufacturing or construction. It is here that there has been the most straightforward transfer of individuals from the secondary sector to service industry employment. But the pattern of transfer differed as between:

a) Transfer as *a career development* in which 1) the individual not only transfers himself but also the kind of work that he has done in the past into service industry, and 2) although the move might be occasioned by forces external to the individual, the transfer is not traumatic in impact.

b) Transfer as *an adjustment to employment crisis* in which 1) the individual moves to a different kind of work, often of lower social status, and 2) the move is traumatic in that between leaving usually long term jobs in manufacturing, generally through redundancy, and taking the present job, the individual experiences periods of unemployment, temporary jobs and the like.

Again there are exceptions, a minority of older men do not fit the above patterns in that they have either never worked in the secondary sector or, prior to their present job they have alternated between work in both sectors. We will refer to these as *always service sector* workers and *formerly sectorally mobile* workers.

The rest of this chapter will be concerned to fill out the detail of the following patterns of labour market experience:[1]

Younger workers	
Older women	Post-domestic workers
	Non-domestic workers
Older men	Career developers
	Post-employment crisis workers
	Always service sector workers
	Formerly sectorally mobile workers

Younger workers.

Twelve of the workers in the sample were aged thirty years or less at the time of the survey, nine women and three men. They included six women academic technicians, two women caterers, one female cleaner, one male security officer, one male attendant and one male maintenance technician. Their average age was twenty-four years.

The younger workers' labour market experience has been marked first of all by instability of employment. Of the twelve people here, nine have been unemployed or participated in a Government training scheme since they left school or college. The contrast with workers over thirty years of age is striking, only five of the older fifty-five experienced unemployment up to 25 years of age. Furthermore, of the three younger workers who have been continuously employed, one has in fact been made redundant twice (see below) and one took a temporary job and then worked part-time in a shop after graduating with a degree. So, only one has has had secure employment over the whole time since leaving school or college.

The second point of interest concerns the process of getting the present job. In nine of the twelve cases the job they have now has been obtained by a series of steps by which better employment conditions have gradually been built up, or their present job is hopefully, from the individual's point of view, such a step to better things. It is very difficult to give a general account of this process of 'edging into' the labour market because the steps depend upon the type of job and the contacts the person has. But a brief description of these nine cases will establish the point.

One caterer took a temporary job at a City Council summer cafe whilst still technically at school. Although unemployed for a while after this, she had her foot in the door and obtained a permanent job as a catering worker, although she had to transfer to the Polytechnic when the restaurant she worked in closed down.

One academic technician, a graduate of the Polytechnic, obtained a temporary job employed by a former teacher. This was later made permanent.

One academic technician, again a Polytechnic graduate, is now in her second temporary job at the Polytechnic, covering for a worker on maternity leave.

One academic technician first obtained a term time only job which was later made permanent.

One attendant was placed at the Students' Union during his work experience scheme and was asked to stay on temporarily when the scheme finished. He was thus in a position to 'hear' of his present job.

One cleaner has spent 18 months unemployed (in three spells) since 1981 and is prepared to do anything as an alternative. So she works part-time.

One caterer still hopes to get promotion to a cooking job from her work as a cashier (formerly a general assistant). She has taken a catering course and prepares pub food at the weekend.

One academic technician did two spells as a temporary clerk in the Polytechnic, at the end of the second of which she was offered the opportunity to apply for her present post.

One maintenance technician began his apprenticeship in the motor components industry but was made redundant after two years. Fortunately his training officer was able to fix up a job for him with the council where he could transfer his apprenticeship. When he qualified he was again redundant but was again 'spoken for' by his supervisor and moved to the Polytechnic.

The third feature of the younger workers' labour market experience is the brevity of their manufacturing sector employment, displayed in table 5.1.

Table 5.1
Young workers experience of manufacturing.

Time in manufacturing	Time in labour market	Remarks
None	8 years	
None	9 years	
1 year	3 years	Sandwich course
None	2 years	
1 year	7 years	Sandwich course
3 months	7 years	Temporary job
2 years	12 years	Trainee
1 year	4 years	'Cowboy firm'
None	6 years	
None	5 years	
1 month	3 years	'Cowboy firm'
2 years	7 years	Redundant

We can see from this, first of all, that the experience of manufacturing industry is rather slight. Five people have never worked in this sector, for the other seven, employment in manufacturing has occupied a small proportion of their working lives. Furthermore, in every case of manufacturing experience some 'remark' is called for as indicated in the right hand column. Two people gained their manufacturing experience whilst students on sandwich courses and so had not actually entered the labour market proper. One person's job was temporary, another two (marked on the table as 'cowboy firms' offered what the two people regarded as unacceptably low wages[2] in dirty, cold conditions. In both their opinions their employers used Government employment schemes to exploit young workers. One started an apprenticeship but was made redundant. The case marked 'trainee' in the table is something of an exception. She took a job training to be a jeweller, liked the work and so did a full-time course before spending a year in jewellery manufacture. Her reasons for leaving the manufacturing sector were to do with the location of the family home.

Older women workers, post-domestic break.

Twenty-one women workers are included in this group, their ages ranged from thirty-one to fifty-nine with the average being forty-two. They included eleven of the women cleaners, nine of the caterers and one woman technician.

After leaving the education system these women entered a period of employment which often covered work in both manufacturing and service industries. This applied in sixteen of the cases, in the other five, two worked only in manufacturing and three only in services. Typically they had a number of jobs, each of which lasted for a relatively short time, before leaving the labour market for unpaid domestic work. The number of years spent in these 'prehousewife' jobs is shown in table 5.2, broken down by broad industrial sector.

As mentioned, this period of employment comes to a close when the women left the labour market for domestic work. As table 5.3 shows this generally occurred in the woman's early twenties , the average age at which the twenty women included in the table first began unpaid domestic work was twenty-one. The table also shows a strong association between this and the birth of the woman's first child. In fourteen of the cases unpaid domestic work begins a few months prior to the first child's birth. In three further cases domestic work is associated with marriage rather than the first child but in each of the three, marriage entailed geographical movement on the woman's part. In one case the woman continued paid employment until the birth of her second child. In another, the birth of the first child coincided with the end of the marital relationship, so that the woman had to carry on working. Conversely, one woman gave up employment because she believed she was pregnant but this turned out to be a 'false alarm'. She remained at home and her first child was born the next year.

Altogether these women had spent thirty-four 'work periods' in unpaid domestic work. In twenty-eight of these the reason given for not being in paid employment at the time was concerned with either pregnancy or child care. In four cases some other domestic reason was given, usually the care of another relative. In one case the woman was made redundant from her former job, in only one case did the woman choose not to work for a while for some other reason.

There is considerable variation within this group as to the length of the absence from paid employment, its link to marital circumstances, and the age of children and whether the 'domestic interlude' is continuous or not. All that is relevant here is the simple point that domestic work forms a break between employment after leaving school/college and a later period in the labour market. Prior to the 'domestic interlude' these women (i.e. the twenty for whom we have sufficient data) had sixty-nine periods of paid employment. Of these, sixty-six had been full-time, only three part-time. Thirty-six had been spent in the secondary sector, thirty-three in service industries. After the 'domestic interlude' the women had had eighty paid jobs. Only twenty of them had been full-time, sixty part-time. Furthermore, the bulk of these jobs had been in the service sector, sixty-five, with only fourteen in the secondary sector and one an extended period of casual employment across both sectors.

Table 5.2
Number of years (*) per job between entering labour market and leaving for domestic work. By industrial sector.

Manufacturing	5				
Rest (**)					
Manufacturing	1	1			
Rest	2	1	1		
Manufacturing	4				
Rest	1				
Manufacturing	4				
Rest	7				
Manufacturing	1				
Rest	0	0			
Manufacturing	1	5			
Rest	2				
Manufacturing	1	2	0	2	
Rest	1				
Manufacturing	3				
Rest	1				
Manufacturing	4	0	0	0	
Rest	0	0	0		
Manufacturing	0				
Rest	6				
Manufacturing	0	0	0	0	1
Rest	1	0	0		
Manufacturing	3				
Rest	4				
Manufacturing	1	0			
Rest	1	2			
Manufacturing	2				
Rest	4				
Manufacturing	0	2	2		
Rest	1	2			
Manufacturing	5				
Rest					
Manufacturing					
Rest	1	2	1	2	1
Manufacturing	0	2	4	1	1
Rest	0				
Manufacturing					
Rest	2	3	5		
Manufacturing					
Rest	4	0			

* In the table 0 years means a job which lasted less than 6 months.

** In the table 'Rest' refers in all cases but one to a job in the service sector. The exception is one woman who spent one year in military service. Also, the table covers only 20 of the 21 cases because for one woman we were not able to construct a detailed work history for reasons explained in chapter 2. These largely concern the sheer multiplicity of jobs she held (in both manufacturing and services) so that it is ironic that she is excluded from this table.

Table 5.3
Leaving employment for unpaid domestic work. Age and birth of first child. Post-domestic women.

Age	Month and year when first entered domestic work	Month and year when first child born	Remarks
23	9 67	12 67	
21	63	10 63	
19	45	9 45	
21	3 54	6 54	
21	11 71	1 72	
19	4 66	9 66	
22	11 58	2 59	
21	10 66	11 66	
21	2 72	6 72	
19	2 65	5 65	
25	5 69	7 69	
22	4 57	8 57	
20	4 65	9 65	
19	3 73	3 73	
20	7 67	11 68	Married and moved.
26	6 75	None	Married and moved.
17	9 66	10 67	Married and moved.
21	11 58	1 57	Birth of second child.
22	76	1 73	Abandoned by husband.
22	53	10 54	'False alarm'.

One case is missing due to insufficient data.

Older women workers, non-domestic interlude.

Only two of the women in the sample aged over thirty had not spent some time in unpaid domestic work. One was a cleaner, the other an academic technician. The cleaner, now sixty-one years of age, had worked most of her life in various residential domestic jobs, the technician, now forty years old, had worked in local government prior to the Polytechnic. We will keep these two people separate from the group above to see if there is any systematic difference between them when we turn to the job estimations and comparisons in chapters six to ten.

Older male workers, transferring to services as a career development.

In chapter four we have established that the male workers at the Polytechnic had spent the great majority of the working lives prior to their Polytechnic employment in the secondary industrial sector. (See table 4.10). This applies particularly to the older men as we have noted that younger workers have very little experience of secondary sector employment. Thus for the majority of older, male workers the significant feature of their pattern of labour market history relevant here is their move from the secondary sector to service sector employment. The first group of men to be considered made this transfer as a 'career development'. The transfer process in this sense has three component characteristics. First, prior to the last entry into service sector employment the individual had spent the predominant part of his working life in secondary sector employment. There is thus a clear move from one sector to another. Second, the transfer process can be called a 'career development' because during working life the individual builds up skills in a particular type of work which are at least maintained, in other cases enhanced, in the move to service industry. Third, the shift to service industry might be actually occasioned by the impingement of external economic factors on the individual but these do not have a traumatic impact on him. He is able to respond to such circumstances in a way which is to his benefit.

Eleven men have a pattern of work history that can be characterised in this way, seven academic and four maintenance technicians. Their average age was forty-seven.

In all cases, prior to the point of transfer to service industry, a large proportion of these men's paid, civilian work[3] had been done in the secondary sector.[4] In fact, in eight of the eleven cases, prior to the point of transfer, all paid, civilian working life had been in secondary sector employment. The remaining three had spent 80%, 72% and 63% of their pre-transfer working lives in manufacturing or construction.

Whilst working in secondary industry these people acquired skills and experience that they were able to carry over into the service sector at their point of transfer. Various indicators of this are shown in table 5.4; their major occupations prior to transfer and those after, how their Socio-Economic Group changed with the transfer, and whether they had been apprenticed or obtained some other qualification prior to the transfer.

As can be seen from table 5.4 all but two of these men completed apprenticeships. In fact, the two who did not are exceptional in that they entered the labour market during the Second World War when apprenticeship training was curtailed. In peacetime conditions it is likely that they would have completed apprenticeships, that being their own view. Five of them had obtained some post-school qualification (other than that connected to apprenticeship). Whilst in secondary industry they had worked in skilled manual occupations, as ancillaries to professionals or as a professional himself in one case. After moving to services they obtained occupations which were, if not exactly the same kind of work, at least extensions, in other cases enhancements, of the kind of work they had become experienced and skilled in. So in each case their position in the Socio-

Table 5.4
Indicators of transfer as 'career development'.

Apprenticeship		Other	Main occupations		S.E.G.	
Started	Completed	post-school qualifications	Before transfer	After transfer	Before	After
Yes	Yes	Yes	Toolmaker	Lab. technician	9	5.1
Yes	Yes	No	Electrician	Electrician	9	9
Yes	Yes	Yes	Electrician	Research tech.	9	5.1
No	No	Yes	Metallurgist	Senior technician	4	4
				Lab. administrator		1.2
Yes	No	No	Aircraft fitter			
			Lab. technician	Lab. technician	5.1	5.1
Yes	Yes	Yes	Printer	Printing tech.	9	9
Yes	Yes	Yes	Dental tech.	Design & devel-		
			Draughtsman	opment engineer	6	4
Yes	Yes	No	Electrician	Electrician	9	9
Yes	Yes	No	Plumber	Plumber	9	9
Yes	Yes	No	Heating engineer	Heating engineer	8	9
				Maintenance superintendent		1.2
Yes	Yes	No	Painter	Painter	9	9

Key to S.E.G. numbers: 1.2=Manager, large establishment; 8=Manual foreman; 4=Employed professional; 5.1=Ancillary to professionals; 6=Junior non-manual; 9=Skilled manual.

Economic Group hierarchy was at least maintained, in other cases improved.

Finally, let us look at the process of transfer itself. Some details relating to this point in these mens' lives are given in table 5.5. Looking first at the mens' ages, eight of them moved between their late twenties and early forties, that is, at a time in their lives when they had accumulated experience but were still in the career advancement phase. One is a little younger (23), but out of his apprenticeship. Two are a little older (49 and 47), clearly experienced and with still a substantial portion of working life to go. The year in which the transfer took place is also significant. It is the case that three of the men transferred in the early 1980's when the great shakeout of labour in manufacturing and construction was occurring. But we can see that in only one of these cases was the transfer directly a result of being made redundant. However, although these shifts to service sector employment are not generally a specific result of the early 80's recession, in six of the cases the men were either made redundant, anticipated redundancy, or had to make a move because of a failing business. Thus external economic forces did prompt their move to service industry. The main point to note here is that in only one case did this lead to unemployment. Faced with loss of employment these men were able to quickly find alternative, and as we have seen, comparable or better employment. Even the one who did become unemployed was only jobless for one month. No doubt to the individuals concerned in the process of redundancy and finding new employment the transfer was a worrying time but compared to the people we will consider next the process was not a traumatic one. These men were in a position to make the transfer from the secondary to the service sector in a relatively smooth and advantageous way.

Table 5.5
Features of the 'career development' transfer.

Age when transferred	Year when transferred	Why left?	Unemployed
27	1980	Move to a better job	No
31	1982	For a more secure job	No No
31	1983	Redundancy	No
37	1966	Redundancy	Yes
44	1968	Redundancy	No
39	1976	Job being deskilled	No
49	1971	Anticipation of redundancy	No
23	1966	Better wages	No
47	1978	Self-employed business failing	No
28	1970	To see more of children	No
28	1959	Redundancy	No

Older male workers, transferring to services as an 'adjustment to employment crisis'.

We now come to a second form of sectoral transfer which is very different from the first. However, one feature is common to both, for here again 'transfer' means a move to the service sector after the majority of previous working life (again excluding military service and unpaid work) has been spent in secondary industry. In this case though the transfer does not involve the carry over of experience accumulated in the secondary context. Rather, people move to different kinds of work, often of lower status than they had been doing in their manufacturing sector jobs. Finally, the process of transfer itself is marked by trauma. The period of transfer is extended, includes one or more spells of unemployment, temporary employment, working for 'cowboy' firms[5] and retraining, sometimes of peripheral relevance. This phase in life is often made more distressing by coming late in life after working for the same employer for many years prior to being made redundant.

Let us turn to the people in the sample who arrived in service industry by this path. It applies in twelve cases, six are currently employed as security officers, four as cleaners and two as attendants. Their average age is fifty-four years.

Most of these men had spent the bulk of their working lives in manufacturing industry. In five cases this accounted for all their civilian, paid employment. In a further four cases it accounted for over 90% of their working lives For the rest, employment in manufacturing had not dominated their lives quite to this extent, the proportions of paid, civilian employment in manufacturing being 61%, 58% and 42%. But in each of these three cases the period immediately prior to transfer had not only been spent in manufacturing but this lasted for a considerable span of time; sixteen years, nineteen years and twenty-one years. Thus it is meaningful to categorise them as transferring from manufacturing to services in the same sense as those who had spent more or less all their working lives in that sector.

Table 5.6 gives an indication of the industrial background of this group of workers.[6]

Table 5.6
Indicators of transfer as
'adjustment to employment crisis'.

Apprenticeship		Other post	Main occupations		S.E.G.	
Started	Completed	school qual	Before	After	Before	After
No	No	Yes	Lorry driver	Porter	8	11
			Storekeeper	Security		10
No	No	No	Engine fitter	Lab. attendant	9	9
Yes	Yes	No	Typesetter	Security	9	10
Yes	No	Yes	Fire officer	Security	5.2	5.2
Yes	Yes	Yes	Electrician	Porter	9	10
				Cleaner		11
No	No	No	Chemical etcher	Cleaner	9	11
No	No	Yes	Security	Security	2.1	10
						5.2
Yes	No	No	Assembly line fitter	Attendant/caretaker	10	10
No	No	No	Engine fitter	Security	9	10
Yes	No	No	Toolmaker	Cleaner	9	11
No	No	No	Paint rectifier	Cleaner	9	11
Yes	Yes	No	Coach builder	Security	10	10
			Assembly line fitter			

Key to S.E.G. numbers: 1.2=Manager, large establishment; 5.2=Non-manual foreman/supervisor; 8=Manual foreman; 9=Skilled manual; 10=semi-skilled manual; 11=Unskilled manual.

As a group these people are rather less qualified than the 'career developers'. Six of the fourteen had begun apprenticeships but only three had completed them, four of them had obtained some other qualification in their time in manufacturing. But whether qualified or not, they had not accumulated skills and experience which they were able to transfer with them to service industry. In only two cases is the kind of work done in their current job akin to the main occupations they have held in the past. These two are very experienced security and emergency officers who had held senior positions at large manufacturing plants. Although the S.E.G. position of one of these appears the same, he in fact has a far less responsible job now than in the past. This shift down in the social hierarchy is a further feature of the transfer. Three people's current occupations place them in the same S.E.G. group as they attained in manufacturing but nine have moved downwards in the social scale. None have moved upwards.

Turning now to the actual process of transfer. Table 5.7 gives some indicators of when and how this occurred.

Table 5.7
Features of the 'adjustment to employment crisis' transfer

Age left manufacturing	Year left manufacturing	Years with last firm	Employer	Why left?	Unemployed
57	1979	16	Massey-Ferguson	Redundancy	Yes
46	1980	30	B.L. Canley	Redundancy	Yes
28	1982	13	Printing firm	Redundancy	Yes
58	1982	21	B.L. Courthouse Green	Redundancy	Yes
58	1980	40	B.L. Canley	Redundancy	Yes
36	1970	14	Dunlop	Redundancy	Yes
55	1980	14	Chrysler Stoke	Redundancy	Yes
38	1979	14	B.L. Canley	Redundancy	Yes
50	1980	30	B.L. Canley	Redundancy	Yes
55	1981	10	B.L. Canley	Redundancy	Yes
55	1980	21	B.L. Canley	Redundancy	Yes
44	1980	17	B.L. Canley	Redundancy	Yes

It is clear that the people in this group moved into services as a result of the recession of the early 1980's. All but one of them left manufacturing industry between 1979 and 1982[7], all of them as a result of redundancy, largely from vehicle manufacture (ten of the twelve). All of them experienced a period of unemployment at this time. However this bald statement hardly captures the difficulties which redundancy brought. Table 5.8 lists the events of the period of transfer from leaving steady employment in manufacture to working at the Polytechnic.

Table 5.8
Events in the transfer from manufacturing to services.

Sick 9 months...unemployed 3 months.

Unemployed 2 years.

Unemployed 4 months...'cowboy' engineering firm 5 months... unemployed 1 year 3 months...security firm (7x12 hour shifts per week) 3 months...temporary security work at the Polytechnic for 6 months before offered permanent job.

Temporary security work at former plant 4 months...unemployed 8 months.

Unemployed 2 months...Temporary job 3 months...unemployed 1 year 3 months.

Unemployed 1 year 3 months.

Unemployed 3 months.

Unemployed approx. 9 months...Temporary job 5 months... unemployed 7 months...Training course for welding 6 months... unemployed 2 years 9 months.

Unemployed approx. 2 years...Car park attendant 3 months.

Unemployed 2 years 3 months with a break during this time for a training course in engineering, (4 months).

Unemployed 11 months.

Unemployed 12 months.

I think this list of events makes clear that the process of transfer for these people was not a smooth one but a traumatic adjustment to a crisis of employment. Most of them had formerly worked in what used to regarded as the best plants in the city (particularly 'the Standard', i.e. B.L. Canley, as ex-workers invariably called it).[8] They had generally worked in these plants for long spells of time, an average of twenty years with the employer prior to redundancy. As we can see from table 5.8 this was followed by extended periods of unemployment and other activities before finding a generally different kind of job of rather lower status in the service sector.

There is the possibility that this group of people came to service sector employment at the end of their working life when they were seeking 'lighter' jobs.[9] Clearly this does not apply to the younger members of this group but the fact remains that seven of them were in their fifties at the time they left manufacturing. Nevertheless this is not a plausible interpretation. None of them chose to leave their former manufacturing jobs, they were all forced to leave what had been long standing jobs by compulsory redundancy. Further, during the period of transfer many had taken very demanding jobs for what they regarded as 'cowboy' employers offering poor conditions of employment. This is hardly evidence of a search for lighter work.

Thirty-one of the thirty-four men in the sample were over thirty and we have discussed the work history pattern of twenty-three of these under the two types of transfer from the secondary to the service sector. This leaves eight of the older men who cannot easily be characterised as 'transferring' sectors, four because they moved between the sectors at various points in their lives and four because all their working lives were spent in the service sector.

Older male workers, formerly sectorally mobile.

There are four men here, one cleaner, two protection officers and one maintenance technician, with an average age of 44. Like the women workers we considered earlier, during the earlier part of their working lives these men moved between manufacturing and service industries. Indeed their mobility between jobs is their common characteristic. Unlike the women there is no interruption in this pattern of movement by domestic work but in each case the mobility between jobs ends in the mid-1970's, since when three have been at the Polytechnic and one has had only one job before coming to the Polytechnic in 1979.

The following sketches of these men's careers indicates their mobility between occupations and industries.

One starts work in the early 1950's as an assembly worker in the car industry...becomes a milkman...hospital porter...salvage man in the aircraft industry...foundry moulder in the car industry...before coming to the Polytechnic in 1974.

One starts work in the early 1950's serving his apprenticeship in engineering...does his National Service...becomes a fitter in the electrical engineering industry...emigrates to Australia where he works on a poultry farm and in engineering...returns to Britain and to electrical engineering...goes self-employed in the cardboard reclamation trade...becomes a postman... a hospital porter in 1975...before coming to the Polytechnic in 1979.

One starts work in the late 1950's serving his time as an electrician, which is the occupation he has always done but in a wide variety of industries...electricity supply...higher education...vehicle components...lens manufacture...timber mills and electrical engineering in Canada...H.M. Dockyard... cement manufacture...hotels...before coming to the Polytechnic in 1973.

One starts work in the mid-1960's and for ten years has a succession of jobs (he estimated over 30) in a wide range of occupations and industries...warehouseman in the retail trade...coal mining... labouring work in various engineering factories...the army...cooking in hotels and local authority homes...before coming to the Polytechnic in 1974.

The main question here is why the pattern of mobility between occupations and industries, moving on every few years, tends to change in the mid-70's when three arrive at the Polytechnic, the other having two jobs in the last ten years compared to his nine in the previous twenty-three years? A number of factors have to be included here. In one case the onset of a close personal relationship coincides with the change in job pattern and was seen by the person himself as 'settling' him down. But in the other three cases there is no coincidence with the entering of marriage or the birth of children. Again, it might be that, after trying a variety of jobs, the individuals found work which they found satisfying and did not wish to leave. Here we can turn to the data on jobs estimated as 'best' and 'worst'. Only one man rated his present job particularly highly, saying it was his best job in terms of money, job autonomy and overall. Two of the others rated their present jobs as their best jobs in one respect, the other did not itemise his present job in his comparisons. On the other hand, none of the four identified their present job as their worst job in any of the five respects asked about. Thus this evidence suggests that the men are not dissatisfied with their work but neither are they highly satisfied. As one man put it he would have moved on if he could.

There is a third factor in explaining their switch from job mobility to job stability. This relates to the timing of the change in pattern, during the mid-seventies and the tightening of the labour market. Prior to that time men such as these, two with apprenticed trades under their belt and two highly adaptable and experienced industrial workers, could move around the labour market as they became bored with one job or were attracted to another. Their job stability over the last ten years suggests that this is much decreased.

Older male workers, who have always worked in the service sector.

This leaves four of the men over thirty who have never worked in the secondary sector industries.[10] Two of them are employed as academic technicians, one as a protection officer and one as a cleaner. Their average age is forty-six years. The two technicians have worked in higher education laboratories all their civilian working lives,[11] the protection officer worked in hotels and domestic service before coming to the Polytechnic. The cleaner was employed as a commercial traveller and school caretaker before the Polytechnic.

Conclusion.

We can conclude this discussion by returning to the first question the research project sought to throw light upon; has there been a move by *individuals* from the structurally declining secondary sector to jobs in service industry? We have established that the great majority of the sample had, at some point in their lives, been employed in secondary industry. However, this chapter has shown that the place of such secondary sector experience differs in the working lives of people and it has been necessary to sort out these patterns of participation in the labour market over the course of individuals' working lives.

We found that amongst the younger workers in the sample (i.e. those under thirty years of age) there had been very little transfer from manufacturing to services for the simple reason that these people had had very little employment in the secondary sector since entering the labour market. Amongst the workers over thirty very different patterns of experience are evident for women and men. Most of the women had worked in the secondary sector but this was heavily concentrated in their younger years when there was roughly equal employment in service and manufacturing jobs. Most of the women left paid employment for a 'domestic interlude' usually associated with the birth of their first child. When they returned to work the jobs they took were then predominantly located in the service industries. But as well as this, these 'post-domestic' jobs were also mainly part-time. This would seem to indicate that the shift of these women workers from manufacturing to service sector jobs is largely a function of their life-cycle stage rather than the decline of manufacturing jobs. Women with continuing domestic responsibilities seek part-time employment which is largely located in the service sector. Thus in 1985 9% of all women part-time employees worked in manufacturing, 89% in service industries.[12] Yet the degree of difference between services and manufacturing as regards the availability of part-time employment has not always been as great as this. A consistent series of figures is not available but various sources show that there has been a decline in part-time employment in manufacturing since the 1970's. This in fact reversed the post-war trend. An article in the Department of Employment's "Employment Gazette" in 1973 commented '...since the early 1950's there has been a gradual underlying upwards trend in the proportion of women employees in manufacturing working on a part-time basis' (November 1973, Pp.1091). This indeed was the case. In 1950 11.8% of women in manufacturing worked part-time. By 1968 this had increased to 17.7%[13] This statistic refers to the proportion of women in manufacturing who worked part-time. If we turn to the proportion of all part-time women workers who worked in manufacturing we find that the 1985 figure above is roughly half what it was in the early 1970's. Using 1971 Census of Population data, 18% of women working less than 30 hours per week were employed in manufacturing industries, 74% in services.[14] An article in the "Employment Gazette" in November 1977 put the proportion of part-time women in manufacturing at 13.6% in June 1976[15] By December 1980 the percentage was 10.1% and it seems to have stabilised around 9% from the early 1980's.[16]

From this we can note that the movement of women from manufacturing and services prior to domestic work to service employment after domestic work is not solely a function of life- cycle development. As well as this there has been a decline of part-time employment in manufacturing. The sample data provides some support for this in the cases of women who have worked in manufacturing after their 'domestic interlude'. This applies in six cases but in only one case has this manufacturing employment occurred in the recent past, one woman working as a caterer in the canteens of first a glass and then a machine tool factory. The other five all found post-domestic manufacturing employment prior to 1975, four of them as assembly workers in the electrical engineering industry, one in a variety of light engineering jobs. Before the mid-seventies there was thus a wider range of part-time employment available in terms of industrial sector. Although the move of women from manufacturing to services is largely a function of life-cycle stage it is also influenced by this shift in the location of part-time jobs.[17]

Over three-quarters of the older men in the sample can be said to have transferred from the secondary sector, in that, prior to their employment in the service sector, they had spent most of their working lives in secondary industries. There are, however, two distinct pathways here, the first where men move to service industry as part of a 'career development', which might be occasioned by external economic constraints but is not particularly traumatic for the individual concerned. By contrast another group of men have been forced to leave the secondary sector in traumatic circumstances and move to different work in the services, usually of lower status.

The remaining quarter of the older men had not transferred in either of these senses, in some cases because they had never worked in the secondary sector, in others because they had been mobile between the two sectors throughout their working lives. Yet we noted that such mobility had come to an end in the mid-seventies since when these men had had stable employment in the service sector. In an indirect sense then they may have transferred in that the reduction of job opportunities in manufacturing has curtailed their former mobility between jobs and industries.

So of the people employed at the Polytechnic a large proportion had experience of secondary sector work earlier in their lives. But in only a minority of cases does this mean that people have shifted from secondary sector employment as a direct result of the decline in job opportunities in manufacturing and construction. This applies to our 'adjustment to employment crisis' group. Some of the 'career developers' had had to leave the secondary sector but this move has not been concentrated in the period of rapid job loss in secondary industries but spread over the last twenty years or so. On the other hand the decline in secondary employment had influenced the opportunities available for other workers in the sample. A small number of male workers had been mobile between jobs in both the secondary and tertiary sectors but this mobility comes to an end when manufacturing jobs are no longer available in the mid-seventies and these men settled into stable employment in the services. For older women seeking part-time employment after a break from employment associated with family formation the location of such jobs has been increasingly restricted to service industries. Finally, for younger workers there has been little opportunity for employment in the manufacturing sector during their difficult years establishing themselves in the labour market.

We will now move from establishing the samples' work history to employing the scheme of labour market participation we have developed here in the analysis of people's estimations of their 'best' and 'worst' jobs and how they feel that their present jobs compare to such nominations.

Notes.

1. There is one person who does not fit into this scheme as her part-time job at the Polytechnic is a second job in addition to full-time employment elsewhere.
2. At 1985 values their pay was £28 for a 40 hour week and £30 for a 43 hour week.
3. Three men had had been conscripted for war and national service and one had been unemployed for 6 months, another for 1 month. These periods are excluded from the percentages which immediately follow.
4. In calculating the proportion of time spent in construction and just when transfer from construction to services occurred we have a difficulty here in the classification system employed. The Labour Force Survey scheme we use here places people working for local authority direct labour departments in the construction sector but people working in other branches of local government in 'other services'. A man moving from the Council's direct labour force to the Polytechnic thus moves from one sector to another even though he continues with the same employer. This seems misleading so that for the purposes of these remarks employment in local government counts as service sector employment and the point of transfer is identified as the last move to such service employment.
5. This is a colloquial expression often used by our respondents to describe firms offering employment in very poor working conditions, on exploitative terms, often producing a product or service of dubious quality.
6. To be clear on a couple of points. The 'qualifications' column indicates whether post-school and non-apprenticeship qualifications were obtained prior to the departure from manufacturing. In the 'S.E.G.' columns, 'before' means the social position held immediately prior to departure from manufacturing, and 'after' indicates positions held since moving into permanent employment in service industry. The S.E.G. positions of short-term service jobs held in the period of transition (see below) are ignored.
7. The exception left manufacturing in 1970, a victim of the rundown in the Coventry aerospace industry in the late 60's.
8. The Canley plant was, in its heyday, commonly regarded as the best plant in the City in a number of respects, notably wage levels, shop-floor organisation and worker autonomy. As Tolliday (1986 Pp. 208) notes the plant has attracted a 'fair degree of myth and idealisation', see the work of S. Melman (1957), D. Rayton (1972) and A.L. Friedman (1977). Nevertheless, the length of time our respondents spent at the plant indicates that the mythology has some validity. We will be taking this further in subsequent chapters.
9. Some men clearly would prefer 'lighter' work as they become older, see the study by P. Lyon (1987) of older workers in a paper mill and direct labour building department. The crucial point is that 'lighter' work was available to some extent at these workplaces but *with the same employer*, there is no evidence that people leave a job to risk their chances on the labour market seeking 'lighter' work.
10. This is not strictly the case, one man, now 63 years old had worked in manufacturing for two years, but this was before the Second World War and after that all his civilian employment was in services so that it seemed best to place him here.
11. One began his career by serving in the Navy.
12. **Employment Gazette** December 1985, Labour Market Data Table 1.4
13. Dept. of Employment and Productivity **British Labour Statistics Historical Abstract 1886-1968** H.M.S.O. Table 142.

14. Census of Population, Great Britain, 1971 **Economic Activity Report Part IV** table 5.26. Manufacturing means 1968 Standard Industrial Classification Orders III and V to XIX, Services means S.I.C. Orders XXII to XXVII.

15. 'Annual Census of Employment results for June 1976'.

16. The **Employment Gazette** has published a consistent series in its Labour Market Data sections since 1981.

17. This is discussed further by C.A. Chesterman (1978) and V. Beechey and T. Perkins (1981).

6 Best and worst jobs in monetary terms

The first objective of the research project was to investigate the work histories of people currently working in a service industry, with particular reference to the degree and nature of any transfer by individuals from employment in manufacturing industry. The second objective was to examine the job estimations and comparisons made by the people in the sample. Each individual was asked to identify his or her best and worst jobs in five respects. These were money, skill, job autonomy, trade unionism and the best and worst jobs 'all things considered'. Insofar as the present job was not identified as being best or worst in a particular respect the individual was then asked to compare the nominated job to the current job on a three point scale.

We turn to these job estimations and comparisons now and in the subsequent four chapters, each of which will consider one of the five aspects. In each chapter we will begin with relevant methodological considerations which bear on the quality of the data employed. We will then turn to the distributions of best and worst jobs in terms of the industries in which they are located, and the sex and current occupation of the respondent. A recurring theme will be the extent to which a contrast can be drawn between jobs located in the secondary sector and those in service industries. We will find that in all five respects no such simple contrasts are supported by the data available. However, when we deploy the more sophisticated scheme of patterns of labour market participation developed in chapter five, the selections made and the comparisons drawn show a clear relationship to the individual's biography of work over time.

This discussion begins with the jobs which were selected as best and worst in monetary terms. The questions employed to explore this issue were:

Simply in terms of the money you earned which is the best job you have ever had?

Do your present wages compare VERY BADLY/BADLY or are they NOT MUCH DIFFERENT to that job?

Again simply in terms of the money you earned, which is the worst job you have ever had?

Do your present wages compare VERY WELL/WELL or are they NOT MUCH DIFFERENT to that job?

In general most people were able to identify best and worst jobs in this respect, in fact only one person declined to give an answer to these questions. However, a number of points regarding the comparability of money earnings over time should be noted, beginning with the consequences of the inflation of money values. Clearly, in terms of the amount of money earned as 'the pound in your pocket', recent earnings were impressive as compared to the past. But most people took this point and were comparing the real value of wages. In the interview situation we were able to clarify this quite easily if there was difficulty. Second, although the question asked about the earnings from the individual's job, some people thought in terms of when they were best or worst off in more general terms, taking into account, for example, household income and liabilities. Again this was easily clarified if there was uncertainty. Finally, asking an individual to compare the relative rewards of jobs over time tends to obscure the effects of more generalised increases in the population's standard of living. For example, the contemporary living standards of a security officer might well be better than those experienced by a car production worker twenty-five years ago, even though at that time the car worker was relatively well paid. So the same man holding these two jobs during his lifetime tends to identify the present job as best in money terms even though at the time of being a car worker he was well paid. It is as if the aggregate rise in living standards over the course of a working life gives a boost to the present and downplays relative financial advantages in the past. This was very difficult to clarify in the interview situation.

As mentioned, in most respects any uncertainty as to the meaning of these questions could be overcome in the interview but for the reasons offered above if there is a bias in the form of the questions it was toward favouring present over past incomes.

Turning now to which jobs were identified as best or worst in money terms by the industrial sector in which they were located. In some ways we might expect a contrast here between best jobs situated in the secondary sector and worst jobs in services. We have established that the great majority of the sample had worked in Coventry's secondary industries in the days of the City's prosperity as a manufacturing centre. At that time manual worker wage rates in the City were relatively high due to the operation of the Coventry Toolroom Agreement and the local piece-rate bargaining system.[1] Brown (1971: Pp. 7) supplies data showing the Coventry Toolroom Rate as consistently around £5 per week higher than national average weekly earnings in vehicles and engineering over a fifteen year period. On the other hand, the pay levels of manual workers in higher education are not known for their generosity. In a recent survey (February 1986) of manual workers' pay in universities, 88% had take-home pay less than the current definition of earnings below the poverty level. (NALGO News 1986: Pp. 4-5). However, the jobs selected as best and worst for money by the sample members do not show such a simple pattern These distributions are displayed in tables 6.1 and 6.2.

Table 6.1
Best jobs for money by industrial sector.

Sector	Number	Percent
Manufacturing	24	35.8
Distribution	3	4.5
Other Services	1	1.5
Polytechnic	39	58.2
Total	67	100

Table 6.2
Worst jobs for money by industrial sector.

Sector	Number	Percent
Manufacturing	20	30.3
Construction	6	9.1
Distribution	17	25.8
Commerce	1	1.5
Other services	10	15.2
Polytechnic	10	15.2
Military service	1	1.5
Casual work	1	1.5
Total	66	100

One person did not identify a worst job for money.

What is immediately noticeable from table 6.1 is the high proportion of people nominating a job at the Polytechnic as their best in money terms, almost three out of five made this choice. Insofar as the Polytechnic was not nominated people chose a job in the manufacturing sector. Only four people selected a job in any of the other industrial sectors.

The sectors in which worst jobs for money were located are less clear cut. Only a minority of people selected a job at the Polytechnic but the remaining service sector industries account for over forty percent of the total. Nevertheless this still leaves thirty percent of worst jobs in the manufacturing sector, almost the same proportion as were nominated as best jobs.

To take this further we next turn to which industrial sector is selected in terms of the sex and current occupation of the respondent.

Looking first at the choices of best job for money made by men and women, two-thirds of the women in the sample selected a job at the Polytechnic. This is rather higher than the men but the fact remains that half the men chose a Polytechnic job. Conversely, half

Table 6.3
Best jobs for money. Industrial
sector by sex and occupation.

Sex and occupation	Manufacturing	Distribution	Other services	Polytechnic	Total
All women	8	3		22	33
All men	16		1	17	34
Women acad. tech.	1			7	8
Men acad. tech.	6			4	10
Maint. tech.				5	5
Protection	4		1	5	10
Women cleaners	2	2		9	13
Men cleaners	4			2	6
Caterers	5	1		6	12
Attendants	2			1	3

the men selected a job in the manufacturing sector as their best for money with only a quarter of the women making this choice.

Adding the current occupation of the respondent reveals that, amongst the women workers, almost all the technicians and a large proportion of the cleaners nominated the Polytechnic as the location of their best job for money, with the caterers being less prominent here and selecting a rather higher number of manufacturing jobs. Of the men, all the maintenance technicians chose a Polytechnic job, half the protection officers did so but in the other occupational groups only a minority of each selected a Polytechnic job, slight majorities opting for manufacturing as the location of their best job in this respect.

In table 6.4 the sectoral locations of jobs selected as worst in money terms are presented in terms of the sex and occupational group of the respondent. Only one woman chose a Polytechnic job as her worst in these terms, jobs selected being located in manufacturing (12) and distribution / other services (9 each). Although only a quarter of the men nominated a Polytechnic job as their worst for money, this is markedly more than the women. The rest of the men chose jobs in manufacturing, construction and distribution in roughly equal proportion.

Going on to occupational groups, the women technicians chose a relatively large number of jobs in manufacturing as their worst for money, as did their male colleagues although here four of the ten chose a job at the Polytechnic. The maintenance technicians all opted for jobs in the secondary sector whilst the protection officers' worst jobs for money are located predominantly in distributive industry. Women cleaners chose jobs in manufacturing and distribution as their worst, their male colleagues nominating service sector jobs, three of the six being at the Polytechnic. The caterers' nominations are predominantly non-Polytechnic service sector jobs, the attendants spread through the services.

Although there are many variations in detail here the point that stands out is how favourably the Polytechnic is regarded from the point of view of the selection of both best and worst jobs for money. We have noted that the form of the questions used here might contribute to this but this was not so prominent that it could account for almost sixty percent of the sample defining a job at the Polytechnic as their best, with only

Table 6.4
Worst jobs for money. Industrial sector by sex and occupation.

Sex and occupation	Man*	Const	Dist	Comm	O.S.	Pol	Mil	Cas	Total
All women	12		9		9	1		1	32
All men	8	6	8	1	1	9	1		34
Women acad. tech	5				2	1			8
Men acad. tech.	5	1				4			10
Maint. tech.	2	3							5
Protection	1		6	1		1	1		10
Women cleaners	5		5		2			1	13
Men cleaners		1	1		1	3			6
Caterers	2		4		5				11
Attendants		1	1			1			3

One person did not identify a worst job for money.

* Key to industrial sectors: Man = manufacturing, const = construction, Dist = distribution, Comm = commerce, O.S. = other services, Pol = polytechnics, Mil = military service, Cas = casual work.

These are defined in chapter 4.

fifteen percent selecting a Polytechnic job as their worst.

By far the simplest possible explanation for the high regard for the Polytechnic from this financial point of view is that the wages paid are high relative to wages people have received in other jobs they have held. As the questionnaire asked for earnings in each of the periods of paid employment an individual had throughout working life, we are able to test this explanation by comparing relative earning levels. However before proceeding along these lines a number of methodological points concerning our data on wages and salaries should be noted.

A number of problems concerning the validity of the information supplied on earnings have been documented in chapter 2. There are others concerning the comparability of the data which we can itemise now. The information supplied to us was translated into 1985 values by using an index of purchasing power. However, the data appeared either as gross, annual income or as net, weekly income, depending on whether the job was paid on a monthly salary or weekly wage basis. To make these comparable one had to be adjusted. To make the next stage (see below) more feasible it was decided to convert annual, gross incomes into weekly net incomes. This was done by dividing the annual figure by 52 and to subtract 30% of the gross to allow for tax and other deductions. This is very much a 'guesstimate' as such deductions obviously vary from one individual to another and change over time. It did seem, however, the best that could be done given the available data. Finally, although we now had a figure for weekly, net earnings at constant values, the hours that people worked to achieve this remuneration also varied widely. Hence, as a further step, the net weekly earnings were divided by the weekly hours of work to arrive at an hourly rate. This again added a degree of uncertainty, however, because the hours of work in a given job can either vary at a particular point in time or can change over time. The hours evidence we had took note of such variations in extreme cases but not in a very sophisticated way.

The upshot of all of these points about the validity and comparability of the data on earnings is that it cannot be regarded as being perfectly accurate by any means. However, with due caution, we can explore whether, on the basis of this data the financial favourability of the Polytechnic is borne out by comparison to previous earnings.

To begin with estimations of best paid jobs in comparison to reported wage levels. As noted above, thirty-nine members of the sample identified their present job as their best 'simply in terms of the money earned'. However there is no straightforward correspondence between this judgement and the reported earnings of the sample. In twenty-seven cases, the Polytechnic job was indeed the best paid job, in terms of hourly net rate, that the individual had held. But of these twenty-seven only nineteen selected their Polytechnic job as their best as far as money was concerned, eight others choosing another job which was in fact paid at a lower rate. On the other hand, thirty-eight people were paid at their highest hourly rate in a job prior to their present one.[2] Nevertheless, despite having a better rate earlier in their working life nineteen of these thirty-eight nominated their present job as their best as far as money was concerned.[3]

So, we have relevant data on all but one of the thirty-nine people who chose the Polytechnic as their best job for money. In half the cases this judgement was supported by what the respondents told us about their earnings over time. In the other half, however, the person's selection of their present job as their best was not substantiated by the information they supplied on earlier earnings. Given this and the point mentioned earlier concerning the tendency of present earnings to be regarded as 'inflated', it does not seem that the favourability of jobs at the Polytechnic from a financial point of view can be explained as simply a matter of present high earnings relative to past earnings. Rather, we will submit that the perception of the present as best for money is related to individual's *experience of the labour market over time*. But before exploring this in terms of the patterns of work developed in chapter five we should pursue a similar exercise in comparing jobs identified as worst for money by comparison with reported earnings over individuals' working lives.

As noted above only ten people identified a job at the Polytechnic as their worst paid job. Beyond this only six of these people selected their present job as their worst paid. Again, using the technique outlined above we can explore whether these judgements are consistent with the information supplied to us about earnings rates in jobs over the lifetime. In this case the low propensity to select Polytechnic jobs as worst is supported by the earnings data. In only seven cases was the present rate of earnings the lowest of the individual's working life, that is, in fifty-eight of the sixty-five cases for whom we have relevant data the respondent had earned a lower hourly rate at some point in the past.

Yet the picture is not quite so simple. Of the seven who were currently earning their lowest hourly rate only one in fact selected the present job as his/her worst for money. Five other people who chose their present job as their worst had in fact earned lower rates in earlier jobs according to the information supplied to us. Again of the fifty- eight people who had had worst paid jobs in the past only half (twenty-nine) nominated as their worst job for money the job in which they were paid the lowest rate. Thus, although it seems clear that the Polytechnic does not pay low wages by comparison with other jobs held by the people over their working lives, it again seems to be the case that people's *perception* of their worst job for money is not simply a function of the actual rate of pay. Once again then it seems worth exploring this question in terms of the pattern of people's working life.

We will now turn to an interpretation of people's estimations of their best and worst jobs for money in the light of their work histories. The scheme of different patterns of labour market participation developed in chapter five will be deployed, beginning with the older men, proceeding with the older women and concluding with the younger workers.

The career development work history pattern.

It will be remembered that these men had spent most of their working lives in the secondary sector but had been able to transfer their work skills into the service sector. The transfer from manufacturing or construction to services as a career development implies the expectation that individuals would not materially suffer from the change and thus we would expect people to select their present jobs as their best in money terms. Insofar as they did not we would then expect them to judge their present earnings as 'not much different' to the job selected as best financially. However, there is a complication here, for although the move to services was, relatively speaking, not traumatic in impact, it was in some cases brought about by 'external forces'. If this was the case we might expect that the job selected as financially most favourable would be the job the individual had lost under duress. Although the transfer had developed the person's career in some respects, it had involved a financial loss.

On the other hand we would not expect these people to select their present job as their worst in money terms. As these are people who have been able to develop a career on the basis of training and qualifications in a specialised form of work it would seem likely that their worst jobs, financially speaking, would be held early in life when they were building up skill in their field. As compared to such jobs their present positions should compare very favourably.

Let us turn to how far these expectations on the relationship between work history and financial estimations and comparisons are supported by the eleven cases who fell into the career developers group. Table 6.5 presents material referring to the best jobs selection.

Six of the eleven selected their present job as their best in money terms. Of the other five, the expectation that their present jobs would not be substantially different to their best jobs is not supported, clearly the move to services has been felt by these people. However, this is consistent with the hypothesis in that four of the five had to leave manufacturing for some reason external to themselves. Whilst they were able to transfer the kind of work they did, this has not always been associated with maintaining earning power. One case, number 11 in table 6.5, does not fit the hypothesis, he does not select his present job as best yet was not required to leave his last job in manufacturing. We return to him in a moment.

The data relevant to the job selected as worst in financial terms is presented in table 6.6. As expected the majority (9) did not select their present job as worst in financial terms, rather their current job compared very well with their worst paid job which was generally held early in their working life.[4] However, there are two cases where the expectation does not hold up, where these career developers feel to be worst off in their present jobs in financial terms. One had been a printer in the manufacturing sector. In terms of the expectations outlined above one might have expected him to have identified his apprenticeship as his worst job for money but he did not. This may be because he could not remember his pay when his apprenticeship ended (which was 30 years before!) but might also be because when he left printing due to being deskilled he was earning very much better money than his current income. At 1985 values his net previous earnings were £267 per week at a rate of £6.67 per hour as compared to his current net earnings of £143 at £3.88 per hour. So the contrast is marked. The second case, number 11 in

Table 6.5
Best job in money terms.
(Career developers)

No.	Selected present job as best money	If no: How present job compares	Reason for leaving last manuf. job
1	Yes		
2	Yes		
3	Yes		
4	Yes		
5	Yes		
6	Yes		
7	No	Not much diff.	Redundancy
8	No	Very badly	Redundancy
9	No	Missing	Redundancy
10	No	Badly	Deskilled
11	No	Badly	Better job

Note: the numbers are simply for reference in the text.

Table 6.6
Worst job in terms of money.
(Career developers)

No.	Selected present job as worst	If no: How present job compares	Age when left worst job
1	No	Very well	16
2	No	Very well	21
3	No	Very well	30
4	No	Very well	20
5	No	Very well	23
6	No	Very well	18
7	No	Very well	20
8	No	Very well	15
9	No	Very well	21
10	Yes		
11	Yes		

Note: the numbers are simply for reference in the text.

both tables 5 and 6, had chosen to move to a service industry and had clearly suffered financially. This may be connected to life-cycle stage, as he has three young children, but like the previous case it is also a matter of skill maintenance. Whilst this man's skills were not directly threatened like the printer's, he emphasised the importance to him of being involved in work at the forefront of technological development which the

Polytechnic offered him. It is of note that neither of these men, when asked to identify their least skilled job would entertain the idea that they had ever had such a thing! So in both cases the transfer to service industry had involved a trade-off of skill and financial reward.

In most cases the career developers' selection of jobs in financial terms fits with the hypotheses drawn from the characteristic features of their work history. Six of the eleven selected their current job as their best. Insofar as they did not do this, their best jobs were the ones they had had to leave in manufacturing. Whilst in these cases the transfer to service industry had been accompanied by earnings reductions, their present jobs were not their worst financially. Two cases do not fit this pattern. Here the 'career development' made sense in skill terms but had involved a trade-off of loss of financial advantage in return for maintenance or enhancement of skill.

The employment crisis work history pattern.

These people had had to leave manufacturing as a result of crisis and went through a traumatic period before finding different and sometimes inferior work in services. Thus one would not expect the current job to be nominated as the best they had held. Rather, as these people often occupied jobs for a long period of time prior to the employment crisis, one would expect the last job before the transfer process to be nominated as their financially most rewarding. By reference to that best job, a further expectation would be that present earnings compare unfavourably.

Turning to the job nominated as the worst in money terms the expectation would be that either the present job or a job held during the process of transfer from manufacturing to services, such as a temporary job, would be nominated. In the latter case, or the case of the selection of another job as financially the worst, one would not expect a great deal of difference between the selected and present job on the grounds that people have had to change the work they do and move to an 'inferior' socio-economic position.

Table 6.7 presents information bearing on the selection of the best job for money by those entering services after an employment crisis. Eleven of the twelve cases selected a job other than the present as their most favourable in financial terms. Of these eleven, seven furthermore nominated the job they had held immediately before the process of transfer. The four who did not follow this pattern each indicated a job which, in terms of the information they supplied to us, was not the best paid job they had in fact held.[5] Rather, they seemed to select a job which in overall terms they strongly identified with, thus three of the four selected this job as their best job overall, again three of the four identified a job which they thought of as their 'trade' (one a toolmaker, one a coach-builder, one a fireman). In one case there were pressing personal reasons why the man strongly identified with the job he selected.

The man who did select his current job as his best for money had some justification for doing so in terms of the earnings he had previously commanded. He had had one job in the past with higher net earnings but his present hourly rate was his highest ever.[6] For this man the result of employment crisis has thus not been particularly detrimental in financial terms and it is not inconsistent of him to select his present job as financially best. But for the other eleven we expected an unfavourable comparison between their best earnings and the present. Table 6.7 shows this to be borne out to the extent that five of the eleven thought their present earnings compared 'very badly' to their best, three thought they compared 'badly' and only three felt present earnings were 'not much different'.

Table 6.7
Best job in money terms.
(Post-employment crisis)

No.	Selected present job as best money	If no: How present job compares	If no: Whether job just before transfer
1	No	Very badly	Yes
2	No	Very badly	Yes
3	No	Very badly	Yes
4	No	Badly	Yes
5	No	Badly	Yes
6	No	Not much diff.	Yes
7	No	Not much diff.	Yes
8	No	Very badly	No
9	No	Very badly	No
10	No	Badly	No
11	No	Not much diff.	No
12	Yes		

Note: the numbers are simply for reference in the text.

Table 6.8
Worst job in terms of money.
(Post-employment crisis)

No.	Selected present job as worst	If no How present job compares	Location of worst job
1	Yes		
2	Yes		
3	Yes		
4	Yes		
5	No	Very well	In transfer process.
6	No	Very well	In transfer process.
7	No	Missing	In transfer process.
8	No	Missing	Apprenticeship.
9	No	Very well	First job.
10	No	Well	First job.
11	No	Very well	National service.
12	No	Very well	In manufacturing.

Note: the numbers are simply for reference in the text.

Turning now to the job selected as the worst in money terms, presented in table 6.8. The expectation that the present job or a job held during the transfer process would be identified as the worst for money was borne out in seven of the twelve cases. In a further four a different consideration enters, in that jobs held during the first few years of working life were identified as the worst for money.

When we turn to the degree of difference between present earnings and the earnings in the worst paid job, the expectation that respondents would not detect much difference between the two is not upheld by the data. Unfortunately two cases are missing here but of the six people who made the judgement five felt that their present earnings compared 'very well' to their worst and one thought they compared 'well'. By comparison with the wages paid in jobs held during the transfer process and jobs held whilst young the Polytechnic is thus perceived by these people as paying good wages.

In conclusion, the men who had moved to their present jobs in service industry after an employment crisis in manufacturing did not select their present jobs as their best for money. Rather they tended to select the job held immediately prior to the crisis and in many cases their present wages compared unfavourably. In some cases their present jobs were their worst for money, in others the worst job lay in the transfer process or a job held whilst young. When asked to compare the present wages with these worst jobs the Polytechnic emerged surprisingly favourably.

The formerly mobile work history pattern.

The four people in this category had a work history characterised by mobility between jobs in both manufacturing and service industry, a mobility which, however, came to a close in the mid-1970's as the labour market no longer allowed men to give up one job and move to another with the ease of former years. At this point in time these four men 'settled' in their jobs at the Polytechnic. It might be expected then that they would locate their best jobs for money in their period of mobility when they could 'play the market' and seek out well paid jobs. Furthermore, given the character of the Coventry labour market at the time these men were mobile, i.e. offering high wages in manufacturing industry, one would expect their best job for money to be located in the manufacturing sector.

As noted this mobility feature ended with their settlement in stable employment at the Polytechnic. As they no longer have the freedom of movement of earlier days, it is feasible that their present job might be their worst for money. However, an alternative is also present. The 'other side of the coin' of movement between jobs is that sometimes a person enters a poorly paid job, by mistake as it were. The mistake can be rectified but playing the market has its downs as well as its ups.

The data on worst jobs is more straightforward and can be considered first. None of the four cases here selected their present job as their worst for money, all selected a job which offered poor earnings in generally unfavourable circumstances, three of the four thus selected the same job as not only their worst for money but also their worst 'all things considered'[7] Again none of these jobs lasted more than one year, the men rectified the mistake and moved on. The jobs concerned were:

1) One man's first job, in a small engineering workshop, using the most basic production techniques in very primitive conditions.

2) One man tried his hand as a milk-roundsman. The other milkmen, for reasons unknown to us, took a dislike to him and not only refused to show him the ropes but sabotaged his work.

3) One man went self-employed for a brief time collecting waste cardboard. The shortage of waste paper did not last long and the business was never viable.

4) One man emigrated to Canada but was unable to settle and returned to the Coventry area. His worst job for money was the one he took immediately on his return, a period of a few months whilst he explored the market again.

So here the choice of worst job is related to the pattern of work history followed by these men, but not to the stabilisation of employment in service industry. Indeed when comparing their present jobs to these worst jobs for money three said that their present earnings compared 'very well', one said they compared 'well'.

When we turn to best jobs for money, two of the men here selected a past job in the manufacturing sector. In one case we do not have full data, this was a man who had had numerous jobs and we were not able to gather detailed information on each. He was quite clear though that his best paid jobs had been in various Coventry car and engineering factories in the 1960's. The other man reckoned he earned £100 per week net in 1969 as a maintenance electrician. Although this job involved a 70 hour week and although there might be some exaggeration here, these were quite remarkable earnings. £100 per week is equivalent to £526 per week at 1985 values, and that is net income! It is no wonder then that the job is selected as best in money terms.

However, the other two men present a puzzle. Both selected their present job as their best for money and in so doing ignore what, on the basis of the information supplied to us about previous earnings, were better paid jobs in the past. A number of explanations for this were tested against the data, for example, examining whether their present domestic commitments compared favourably with the past, but the data did not support any of these. All that can be said is that whilst these mens' *worst* paid jobs had 'stuck' in their memory, their better paid previous jobs had not.

The always services work history pattern.

The distinguishing feature of the four men included in this group is that they had never worked in the secondary sector, their civilian working lives having been spent in various branches of service industry. Our earlier discussion of the industrial sectors in which best and worst jobs for money lay suggested that by comparison with other service industries in which the sample had held jobs the wages paid at the Polytechnic compare favourably. Thus we might expect these people to select their present earnings as their best. On the other hand wages paid to young workers in services are notoriously low so the our expectation would be that worst jobs for money would be identified as those held early in the working life. In addition we might expect that current remuneration would compare very favourably with these early wage levels.

These expectations are fully fulfilled with respect to the selection of best job for money, all four men chose their present job. Jobs selected as worst for money are itemised in table 6.9.

Two of the four selected a job which they left before the age of twenty as their worst job for money. Case number 3 in table 6.9 is in fact somewhat similar, this was the first job held by the man after pursuing a career in the armed services so that he was adjusting to the civilian labour market. The last case does not fit, but in many ways his whole work history is rather unusual in the Coventry context. Now 63, he spent his prime working years in rather low paid work in the service sector at the time when manual workers in Coventry's manufacturing industries were commanding relatively very high wages. The reason for his resistance to the lure of the factory is not known.

Table 6.9
Worst jobs for money.
(Those always in services).

No.	Age when finished	Occupation	How present compares
1	19	Junior lab technician	Very well
2	19	Chef	Very well
3	33	Design draughtsman	Very well
4	33	Assistant caretaker	Very well

Note: the numbers are simply for reference in the text.

The comparisons these people made between their worst paid jobs and their present jobs are sharp, all four of them thought that their present wages compared 'very well' .

The women with a domestic interlude work history pattern.

The advancement of hypothetical expectations regarding women's judgements on their working lives is extremely hazardous. For one thing, a plurality of factors seem to be at work; the woman's age, the industrial sectors in which she has worked, her experience of full- and part-time work, the place of her earnings in the income of both her family of origin and her family/ies of marriage. Furthermore, putting these factors together can lead to the generation of contradictory hypotheses. We can illustrate these difficulties by taking *age* as a starting point. The low wages of young women workers are well known. So one might expect women workers to identify their worst jobs for money to be those held when they were young. But, on the other hand, young women workers (as we have seen in chapter five) tend to have full-time jobs and to have a greater chance of working in manufacturing industry which often pays higher rates of pay than service sector employment. Again, young women workers are often still single and might be allowed a degree of independence in the disposal of their income by their families of origin. These factors might encourage women workers to see their early jobs as being relatively well paid.

Approaching matters from another tack, more mature women workers tend to receive better rates of pay, but the predisposition to thereby define jobs held when older as best for money might be counterbalanced by the greater chance of having a part-time job, a job in service industry, or by the financial demands of domestic responsibilities.

So the position is very unclear and we should use the data available to us to try to sort out the relative weight and inter-relationship of these factors. Let us begin with jobs selected as best for money by the twenty-one women in this group.

Twelve of the twenty-one chose their present job as their best for money. All but one of these were working part-time. We should also remember that all these women are over thirty years of age. This of course leaves nine women who chose a job other than their present one as their best for money. Data pertinent to these jobs are presented in table 6.10.

A quite clear pattern is evident in table 6.10. The jobs selected are all full-time. They tend to be in the manufacturing sector (six of the nine) and they tend to be held when the woman is over 21 (six of the nine). It is noticeable that in seven of the nine cases the woman thought there was 'not much difference' between her present job, financially

Table 6.10
Jobs selected as best for money
(other than present job).
Post-domestic women.

No.	Industrial sector	FT/PT	Age when finished	How present job compares
1	Manufacturing	FT	54	Not much diff.
2	Distribution	FT	26	Not much diff.
3	Manufacturing	FT	34	Not much diff.
4	Manufacturing	FT	21	Badly
5	Manufacturing	FT	19	Not much diff.
6	Manufacturing	FT	19	Not much diff.
7	Distribution	FT	25	Not much diff.
8	Manufacturing	FT	35	Not much diff.
9	Distribution	FT	47	Badly

Note: the numbers are simply for reference in the text.

speaking, and her best job for money, the remaining two seeing their present jobs as comparing 'badly' rather than 'very badly'. Thus the 'post domestic' women in the sample seem to see part-time work at the Polytechnic as comparable in financial terms to the full-time jobs of mature women in manufacturing industry. It is of note that, of the twelve women who chose the present job as the best in this respect, only four had had full-time jobs since the age of twenty-one and only one of them had had a full-time job in manufacturing since that age. Thus those women who had had full-time, manufacturing jobs when of adult years tended to chose such a job as their best for money but to see little difference in money terms between that job and their present one. Those who had not had such a job tended to chose the present job.

These observations are confirmed when we look at the jobs nominated as the worst for money in table 6.11. A number of points are apparent here, the great majority of these women selected jobs held in their youth as their worst for money. Fifteen of the twenty-one nominated such a job. Furthermore, of these, fourteen chose a full-time job, the exception being exceptional in several respects[8]. Nine of the fifteen worst jobs were located in the service sector, six in manufacturing.

The six women who selected jobs held later in life generally chose service sector jobs (five of the six), mainly part-time (four of the six). None chose their present job or indeed any Polytechnic job. Furthermore the comparisons drawn by these women between their present jobs and their worst paid jobs indicate their favourable attitude, relatively speaking, to their present wage levels. Eleven thought their present job compared 'very well' to their worst paid job, five thought the present job compared 'well' and only four saw 'not much difference'. (One did not know).

To conclude this section, the most important factor at work in the judgements about jobs in money terms is the *age* when the job is held. The majority of worst jobs for money were held when the woman was young, the majority of best jobs for money were held when the woman was of adult years. This factor seems to over-ride the fact that jobs held when young were full-time and seems to apply almost as much to jobs in manufacturing as in the service sector. Few of these women remembered their period of being

Table 6.11
Worst jobs in terms of money.
(Post-domestic women)

No.	Age when left job	FT/PT	Sector	How present job compares.
1	23	FT	Manufacturing	Not much diff.
2	18	FT	Manufacturing	Well
3	16	FT	Distribution	Don't know
4	19	FT	Manufacturing	Very well
5	18	FT	Distribution	Very well
6	17	FT	Manufacturing	Not much diff.
7	16	FT	Distribution	Well
8	21	FT	Manufacturing	Not much diff.
9	15	FT	Manufacturing	Very well
10	16	FT	Other services	Very well
11	*	FT	Distribution	Very well
12	18	FT	Distribution	Well
13	19	FT	Other services	Very well
14	17	FT	Other services	Well
15	21	PT	Distribution	Very well
16	38	PT	Distribution	Very well
17	49	PT	Other services	Very well
18	44		Casual worker	Very well
19	28	PT	Distribution	Very well
20	46	FT	Other services	Very well
21	42	FT	Manufacturing	Not much diff.

Note: the numbers are simply for reference in the text.
* This person selected a number of jobs, all held in her teens.

single workers without domestic responsibilities as affluent times. By comparison with these worst jobs for pay, part-time jobs at the Polytechnic compared very favourably. However this seems to indicate the exceptional character of the Polytechnic as a provider of service sector part-time jobs rather than the favourability of part-time service sector work in general. For one thing, insofar as people did not chose youthful jobs as their worst for money they tended to chose part-time jobs in services other than the Polytechnic and to regard the present job as being very favourable by comparison. Again, a majority of the sample chose the Polytechnic as their best job for pay, the exceptions being those who had had full-time jobs as adults in manufacturing. For these women, part-time jobs at the Polytechnic were not perceived as much different in money terms to their designated best jobs.

Women with a non-domestic interlude work history pattern.

Although only two women fell into this group it seemed worthwhile to investigate whether there was any significant difference between the judgements on jobs they make, not having left the labour market for domestic reasons, as compared to the group above. It will be remembered that one woman works full-time as a technician, the other part-time as a cleaner. On the issue of money there seems very little difference. Both chose as the worst job one held in their youth, full-time in the service sector. Both chose their present job as their best for money. Both women had had full-time jobs after they were

21 but neither of them such jobs in manufacturing. Thus in this respect the two follow the pattern of the other adult women workers.

Younger workers.

The characteristic feature of the younger workers' labour market experience is their difficulty in establishing themselves in secure employment. Most of them had experienced periods of unemployment, training schemes, temporary jobs and the like. Thus we would expect them to identify their present jobs as their best in financial terms, as their present jobs are reasonably secure whilst we would also expect them to select a job held in the process of 'edging into' the labour market as their worst in money terms. With respect to the degree of difference between the selected worst job and the present a number of possibilities are open. First, the current job might compare very well because some of the jobs these people have held offered exploitative rates of pay. But, then again, although the Polytechnic hardly comes into this class, some of these workers are either in rather low paid jobs or feel themselves to be doing work for which they are overqualified. So, a rather mixed picture would not be unexpected here.

In point of fact the data match these expectations almost perfectly. Of the twelve individuals, eleven nominated their present job as their best job for money. The one exception is rather puzzling, she selected a job which, on the information supplied, paid almost the same net amount at the same hourly rate as her present job. As this selected job was temporary one would have expected her to have selected the current job.[9]

In the case of selected worst jobs for money there are no exceptions to the expectations stated above. The jobs selected are listed in table 6.12.

In each case the job selected as the worst for money was one directly concerned with preparing for the labour market, (numbers 3 to 5, 8 and 10), establishing a position in the labour market (numbers 6, 7 and 12) or in temporary or short term positions (numbers 1, 2, 9 and 11). Again, as we expected, there is something of a mix as far as comparison with present wages is concerned although with five people feeling their present jobs compare 'very well' and three 'well' the balance tips toward favouring jobs at the Polytechnic.

Table 6.12
Worst jobs in terms of money.
(Younger workers)

No.	Description of job	How present job compares
1	1st job as temp. caterer	Not much diff.
2	Post army, 1 month, storeman	Very well
3	Sandwich course placement	Not much diff.
4	Sandwich course placement	Well
5	Sandwich course placement	Missing
6	Term time job at Poly	Well
7	Trainee jeweller	Very well
8	Training scheme	Not much diff.
9	'Cowboy' firm	Very well
10	Training scheme	Very well
11	'Cowboy' firm	Well
12	Apprentice	Very well

Note: the numbers are simply for reference in the text.

Other.

There is one other case, a woman whose current job at the Polytechnic is her 'second' job. She nominated her current full-time job as her best for money and did not feel she had ever had a worst job financially speaking.

This chapter has provided an analysis of the jobs selected by sample members as their best and worst, financially speaking. Examination of the data showed that there was no simple pattern of people nominating jobs which, in terms of the information supplied on actual earnings, were indeed the best and worst paid jobs. People's perceptions of the financial rewards of jobs varied from the memory of their actual earnings. Again, such perceptions were not simply related to the industrial sector in which jobs had been located. However, in terms of the overall pattern of working life the jobs selected, and the comparisons made with the present job, did fit the scheme of labour market participation developed in the last chapter.

The most noticeable feature of the estimations of best and worst jobs for money was the favourable position of jobs at the Polytechnic. Two qualifications need note here. A number of methodological considerations suggested that there might be a small degree of bias in favour of the presently held job. Also, we must remember that people were making *relative* judgements about the rewards of jobs they had held, not commenting on the absolute value of their wages or salaries. However, this notwithstanding, the Polytechnic is regarded as a good employer in this respect *by comparison to* other employers.

Notes.

1. Wage rates in Coventry during the post-war decades have been authoritatively analysed by Knowles, K.G.J.C. and Robinson, D. (1969) and Brown, W. (1971). The Coventry Toolroom Rate originated in the Second World War when the City's vehicle and engineering industries were engaged in armaments production. The pressure of production for wartime needs generated a persistent shortage of labour to which management responded by paying high rates of pay to semi-skilled production workers on piece rates. The knock-on effects of this were to draw skilled toolroom workers, whose essential work was paid by the hour and was unsuitable to piece-work determination, into semi-skilled work. To avoid this, leading employers and trade unions adopted the C.T.A. in 1941 which rested on the publication, at quarterly intervals, of the average piece-rates paid in the City's leading firms. This rate was intended to be the minimum paid to skilled, hourly paid toolroom workers, but from the early 1940's to its demise in the early 1970's the published Rate came to have a key place in the whole Coventry wage system (and indeed in the wages paid in the U.K. engineering and vehicle industries). Coventry production workers were paid on piece-rates in which decentralised shop-floor bargaining was a recurrent feature. In this process the current Toolroom Rate acted as the baseline. As any group of workers pushed their rate up from the going Toolroom Rate this then fed back into the next quarterly average which then became the starting point for the next bargaining sequence. As Brown put it: 'In these circumstances earnings rise automatically as a result of the day-to-day

operation of the piece-work system.' (1971: Pp. 25)

2. We have data on past earnings on 65 of the 67 people in the sample so that there is a slight disparity between the data on job estimations and data on actual earnings.
3. Of the remaining 19, 8 selected the job with the highest rate as their best job for money, 11 chose a job other than both their present and highest paid job.
4. The exception is number 3 in table 6.6 who identified a period when he was working as a journeyman for the Council in the late 1950's as his job with worst pay.
5. The respective net weekly income (1985 value) of the job they selected and the best paid job they had held were: 8)£100/£200, 9)£86/£221, 10)£129/£135, 11)£85/£167.
6. His previous net weekly income (1985 value) and hourly rates were: £101/£2.52, £65/£1.63, £134/£1.60, £84/£2.27; present: £101/£2.72.
7. The exception nominated a period of self-employment as his worst for money but it did have its compensations apart from being financially unviable so that it was not his worst overall.
8. She chose a 6 hour per week cleaning job taken when her first child was very young and she had been abandoned by the child's father. She did not take her 'domestic break' from paid employment until later. Given these circumstances the job she chose is in some senses part of the youth labour market, in others a post-domestic break job.
9. The selected job paid £96 net per week at a rate of £2.53 per hour. The current job paid £95.20 net per week at a rate of £2.57.

7 Best and worst jobs in skill terms

The following questions were employed to ascertain people's estimations of their best and worst jobs in skill terms and how their present jobs compared.

In your view which of the jobs you have had demanded the greatest level of skill from you?

How does your present work compare with that job in terms of the skills you use? Does it compare VERY BADLY/BADLY or is it NOT MUCH DIFFERENT?

In your view which of your jobs has demanded the least amount of skill from you?

How does your present work compare with that job in terms of the skills you use? Does it compare VERY WELL/WELL or is it NOT MUCH DIFFERENT?

These questions worked well although there were two circumstances in which people found it difficult to respond in a straightforward way. First, some people were reluctant to compare the different kinds of skills required by different jobs as either better or worse. An example is a person who is now a senior technician with some managerial responsibility but in the past had been a skilled manual craftworker. The dilemma was evaluating managerial and manual skills by some common criterion. A very small number of people felt unable to make such a judgement.

A second difficulty occurred in connection with the 'least skill' question in the cases of craft trained or very well qualified workers. Here the suggestion that their work had not been 'skilled' (rather than more or less skilled as was the intention) was rejected with the claim that they had never had an unskilled job and thus could not name the job which was least demanding in this respect.[1]

What these methodological points indicate is that 'skill' is, in part, a social category attached to certain kinds of work roles. For example, the work of men has historically been defined as skilled in relation to that of women (Phillips, A. and Taylor, B. 1986) and skilled work has often been associated with craft organisation (Thompson, P. 1983: Chapter 4). Given this we might again expect a contrast between industrial sectors, in that skilled jobs might be more likely to be found in secondary industry with its greater predominance of male jobs and craft workers, with less skilled jobs being located in the secondary sector. Once again the Coventry context might be expected to highlight this. Within the local industrial culture the city is regarded as a centre of skilled work in manufacturing industry. A quote from an early 'official guide' can illustrate this:

> 'In output we are not thinking of two or three cities in America, the land of enormous mass production, where cheapness is most desired to swamp the world, but we have in mind Great Britain, with its requirements of endurance and good quality, first class workmanship and materials. In these things there are no serious rivals...[to Coventry as a centre of motor manufacture]'. (Coventry City Council 1930).

There is a degree of irony here in that some historians have argued that it was the absence of craft organisation which initially attracted engineering entrepreneurs to the city (Hinton, J. 1973) and others have argued that craft organisation was eliminated by technological and managerial changes in the inter-war years (Whitson, K. 1979). Yet the mythology attached to firms such as Alfred Herbert in the machine-tool industry remained (McG. Davies 1986) and the Census figures continue to show Coventry as marked by an unusually high proportion of 'skilled workers'[2]. This was because in the context of Coventry's large, mechanised factories shop-floor organisational strength allowed large numbers of 'semi-skilled assembly workers and machinists to earn the skilled *wage rate* and thereby to claim the status of "skilled workers"' (Friedman, A.L. 1977 Pp. 204). As one car worker neatly put it:

> 'In Coventry...you've got two categories of skilled worker. There are the traditional skilled workers - the toolmaker, the machine tool setter, the engineering maintenance worker - as well as certain other craft skills that are still necessary. These are usually people who've served their time, though in some cases they may be upgraded from the lines. Then there are a lot of jobs in the motor industry, like fitting on the car track, which in other parts of the country have been allowed to become semi-skilled whilst here in Coventry we've managed to maintain them as skilled jobs. In fact there's a large number of workers in Coventry who are classed as skilled workers, and who get the skilled rate, even though their "skills" would only take a few days for anyone to pick up.' (Quoted in Thompson, P. 1983 Pp. 107).

Thus despite Coventry's association with large scale, mechanised factory production its manufacturing industry was still linked with skilled work.

We now turn to the industrial location of the sample's selection of best and worst jobs in skill terms. Table 7.2, displaying the location of worst jobs for skill, gives considerable support to a contrast between the secondary and service sector in these terms. 73% of such worst jobs were located in the service industries and only 20% in manufacturing and construction. However, the sectoral location of best jobs for skill is not nearly so

Table 7.1
Best jobs for skill by industrial sector.

Sector	Number	Percent
Manufacturing	26	38.8
Construction	2	3.0
Distribution	9	13.4
Transport	1	1.5
Commerce	1	1.5
Other services	9	13.4
Polytechnic	15	22.4
Military service	3	4.5
Other unpaid	1	1.5
Total	67	100

Table 7.2
Worst jobs for skill by industrial sector.

Sector	Number	Percent
Utilities	1	1.7
Manufacturing	9	15.3
Construction	3	5.1
Distribution	10	16.9
Transport	1	1.7
Commerce	2	3.4
Other services	7	11.9
Polytechnic	26	44.1
Total	59	100

Eight people refused to nominate a worst job for skill, see footnote above.

clear cut. Indeed, 52% of such jobs are located in services and only 42% in the secondary sector.

It is clear from tables 7.1 and 7.2 that the Polytechnic stands in a very different position in skill terms to that we discovered in the last chapter. Only 22% of the sample thought their most skilful job had been at the Polytechnic, whilst 44% located their least skilful job there. It might then be that the sectoral distribution is a function of the particular jobs researched at the Polytechnic. Tables 7.3 and 7.4 show that there is indeed a relationship between the sample members' current occupation and the likelihood of selecting the present job as either best or worst in skill terms.

Table 7.3
Selection of present job as most skilled.
By current occupational group.

Occupation	Yes	No
Academic tech.	11	7
Maintenance tech.	3	2
Security	0	10
Cleaners	1	18
Caterers	1	11
Porters	1	2
Total	17	50

Table 7.4
Selection of present job as least skilled.
By current occupational group.

Occupation	Yes	No
Academic tech.	2	16*
Maintenance tech.	0	5*
Security	1	9
Cleaners	10	9*
Caterers	5	7
Porters	2	1
Total	20	47

*These figures include those who refused to nominate a worst job for skill.

Very few of the security staff, cleaners and caterers selected their present job as their most skilled. The only occupations which did so in any numbers were the technicians. Nevertheless, not all the technicians took this view. Again, very few of the security men or technicians nominated their present job as their least skilled, the only ones making this choice in any number being cleaners and caterers. But not all clèaners and caterers took this view. So although the skill character of the present job clearly plays a part here there is a degree of variation. To throw light on this we will turn again to the relation between jobs selected as most and least skilful and the sample's work histories.

The career development work history pattern.

As in the case of judgements on jobs in terms of their financial rewards the fact that these people have pursued a career path of accumulating skills and experience would lead one to expect that their most skilled job would be their present one and their least skilled a job early in their working life when skills were being learned. A variation on the most skilled expectation would be that this would be located immediately before the transfer to service industries as we noted that some of this group were compelled to leave their last manufacturing job even though they were able to shift their skills to the service sector. Insofar as the present job is not selected as the most skilful one would not expect it to compare unfavourably with the nominated job. On the other hand one would expect the present job to compare very favourably with the least skilful.

The relevant data for discussion of the job selected as the most skilful is presented in table 7.5.

Table 7.5
Best job in terms of skill.
(Career developers)

Number	Location of best job	If not present, Why left?	If not present, How present job compares?
1	Present		
2	Present		
3	Present		
4	Present		
5	Present		
6	Just before transfer	Redundancy	Don't know
7	Just before transfer	Better job	Not much diff.
8	Research fellow Higher educ.	More money	Not much diff.
9	Self-employed plumber	Business failing	Badly
10	Family firm	Business failing	Badly
11	Apprentice	End of apprenticeship	Not much diff.

Note, the numbers are simply for reference in the text.

The above expectations are fulfilled in cases 1 to 6 here. The first five select their present job as their most skilful, the sixth his last job in manufacturing from which he was made redundant. Case number 7 also identified the job just before transfer but he was not forced to leave that job so the expectation that the last job before transfer would be selected if the job was vacated under duress is not fulfilled. In the cases of 8, 9 and 10 we must refer to the individual's particular work history. Number 8 had attained a job in university research which he regarded as his most skilled but he was persuaded to leave when an industrial employer, precisely because of his specialist skills, made him an offer he could not reject, an increase in salary of two and a half times! Number 9 regarded his spell of self-employment as his most skilful job because of its wide range of tasks and responsibilities. But the business declined and he was forced to return to employee status. The family firm which number 10 selected as his most skilful was in fact his own family's firm, owned by an uncle in which number 10 carried considerable responsibility and which he hoped to inherit. Unfortunately this did not materialise and the business folded on the death of his uncle.

Number 11 appears anomalous in identifying his apprenticeship as his most skilful job. It might be that this man conflated his time as an apprentice and his period as a journeyman as he continued to work for the same firm, but this is only speculation.

Those who nominated a job other than their present one as their most skilful generally fulfill the above expectations regarding the degree of difference between their most skilful job and their current one. Numbers 7, 8 and 11 saw little difference between the two, number 6 could not say which indicates that the present did not stand out as markedly worse. The two who said that the skill level of the present job compared 'badly' were the two who nominated jobs which had involved them running a business as well as deploying their trade skills.

Turning now to the jobs chosen as least skilful, listed in table 7.6.

Table 7.6
Worst job in terms of skill.
(Career developers)

Number	Location of worst job	How present job compares
1	Never	
2	Never	
3	Never	
4	Never	
5	Pre-apprentice	Very well
6	Apprentice	Very well
7	Immediately post-apprentice	Very well
8	Immediately post-apprentice	Very well
9	Immediately post-apprentice	Very well
10	Sixth job	Very well
11	Fourth job	Very well

Note, the numbers are simply for reference in the text.

The data on jobs selected as worst for skill largely fulfill the expectations stated above. First of all, cases 1 to 4 refused to nominate a 'least' skilful job on the grounds that all their jobs had been skilled. This is quite consistent with their work history pattern of accumulating skill over time. The next five cases (5 to 9) identified jobs done early in their work biography. It is interesting that only one person selected an apprenticeship *per se* here, though all of them served apprenticeships. Rather, one nominated a short term job held whilst he was waiting to take up his apprenticeship, but three others the jobs they took when they came out of their time and were establishing themselves as journeymen in their respective trades. Cases 10 and 11 do not fit the expectations derived from their work history. They are identified in table 7.6 simply by their work period number because there was nothing about the jobs chosen which fits the approach being explored here. In fact, in terms of the data available there seems little to distinguish why these jobs were selected as the least skilful.

The expectation that a marked difference would be perceived between the job identified as the least skilful and the present job is strongly confirmed. All seven who made the comparison felt that their present jobs compared 'very well' to their least skilful job.

The employment crisis work history pattern.

Given the nature of the experience of transfer to service industry one would not expect the present job to be identified as the most skilled the person has held. Rather, the job from which the person was made redundant as the transfer process commenced would seem to be likely. However, there is another possibility here which connects with the pattern of recruitment to the car industry which, it will be remembered, was the main industry from which these people moved. Car manufacture involved a large number of semi-skilled jobs which were often done by people who had left more skilled occupations for the lure of the car plant and its high wages.[3] Thus it might be that people who worked in semi-skilled occupations prior to transfer would identify an earlier job as their most skilled job. We would, however, expect people who had held skilled jobs prior to transfer to nominate that job as their most skilled. In either case, as the transfer had taken these people into different work, we would expect a high degree of difference between the skill level of the most skilful job and that of the present job.

On the other hand the expectation regarding the least skilled job would be that it would either be the present job or a job held during the process of transfer. Insofar as the latter is the case, little difference between the selected job and the present one would be anticipated.

Turning to our evidence pertaining to the job selected as the person's most skilled, presented in table 7.7. As can be seen, none of these people chose their present job as their most skilled. Six people chose their last job in manufacturing before the transfer process commenced. These all held skilled occupations, cases number 1 to 6. Most of the remaining eight had held non-skilled[4] jobs immediately before transfer but had previously done skilled work and this was identified as the most skilled job. In the cases of numbers 7 and 10 in the table this does not appear to apply in that they held similar positions in their jobs just before transfer to the ones they identify as their most skilled. However, both the men involved here would highlight the especial skills of the job they selected, one stressing his responsibility for the delivery of supplies to and from a major oil depot, the other emphasising the specialist character of *press tool* toolmaking. So all the cases comply with the expectations outlined above in terms of jobs nominated as the most skilled. As for the degree of difference between the selected job and the present job, seven people thought their present job compared 'very badly', three said 'badly' and two thought there was 'not much difference'.[5] Again this is very much in line with the expectations stated above deriving from the work history pattern.

Turning to jobs chosen as the least skilled, five of the twelve people here selected their present job and a further six selected a job which was held during the process of transfer to their present positions. In four of the latter cases the job nominated was their very first at the Polytechnic, held prior to obtaining their current one. Thus it seems that the 'internal labour market' within the Polytechnic gives some opportunity for people to move to better jobs in skill terms. The exception to the expected job selection chose his first job as his worst for skill, working as a teaboy in a small engineering works.

Of the seven people who selected a job other than their present one, three thought their current job compared 'very well' to their least skilled, two said the present compared 'well' and only two thought there 'not much difference'.[6] Thus the expectation that there would be little to chose between the present job and the least skilled is not upheld. As we

Table 7.7
Best job in terms of skill.
(Post-employment crisis)

Number	Best skill last job before transfer	Job before transfer	Most skilled job	How present job compares?
1	Yes	Engine fitter		Not much diff.
2	Yes	Printer		Badly
3	Yes	Electrician		Very badly
4	Yes	Chemical etcher		Very badly
5	Yes	Engine fitter		Not much diff.
6	Yes	Paint rectifier		Very badly
7	No	Foreman storekeeper	Foreman checker oil depot	Very badly
8	No	Security officer	Naval officer	Badly
9	No	Track assembler	Maintenance fitter	Badly
10	No	Toolmaker	Press tool fitter	Very badly
11	No	Track assembler	Coachbuilder	Very badly
12	No	Security officer	Fireman	Very badly

Note, the numbers are simply for reference in the text.

shall see again in this chapter there are many jobs inferior in skill terms to those offered by the Polytechnic.

The formerly mobile work history pattern.

The characteristic feature of these four men's work history is the pattern of mobility between jobs and industries which was curtailed in the mid-1970's. Quite what the implication of this is for skill is rather uncertain. However, three of the four received training in a particular occupation even if they did not consistently pursue that occupation in all their subsequent jobs. Two completed apprenticeships, one as an electrician and one as a fitter in electrical engineering, another was trained as a cook during his time in the army. Thus we might expect these men to select as their most skilful job a time when they were exercising skills they had been trained for. But as they had not pursued a career in their trades, perhaps their present job will not compare too unfavourably.

Turning to their least skilled jobs, again the general impression that the Polytechnic offers more skilful jobs than others on offer in the labour market might suggest that during their varied careers these men will have entered low skill jobs to which their present work compares reasonably favourably.

The data presented in tables 7.8 and 7.9 supports these suggestions. The jobs selected as the most skilful are those in which the men exercised their trades, but the comparison with the present job does not bring out a strongly perceived difference in skill level. On the other hand the jobs selected as least skilful are generally not in the men's trades and the present job compares very favourably.

Table 7.8
Best jobs in terms of skill.
(Post-mobile men)

Number	Year job finished	Industrial sector	Occupation	How present job compares?
1	1967	Manufacturing	Salvage man	Not much diff.
2	pre-74	Distribution	Cook	Badly
3	1965	Manufacturing	Maintenance electrician	Badly
4	1972	Manufacturing	Electrical fitter	Not much diff.

Note, the numbers are simply for reference in the text.

Table 7.9
Worst jobs in terms of skill.
(Post-mobile men)

Number	Year job finished	Industrial sector	Occupation	How present job compares?
1	1961	Other services	Hospital porter	Very well
2	pre-74	Manufacturing	Engineering labourer	Not much diff.
3	1979	Other services	Kitchen porter	Very well
4	1972	Distribution	Service electrician	Very well

Note, the numbers are simply for reference in the text.

The always services work history pattern.

We saw earlier that whilst jobs at the Polytechnic are not generally selected as the most skilled, neither are they generally seen as the least skilled. In the case of these four men, all of whose civilian employment has been in services, we can enquire whether this general pattern is found in this more restricted scope. Tables 7.10 and 7.11 suggest that it is. None of the four selected his present job as his most skilful, though no clear pattern on the degree of difference emerges. On the other hand, only one of the four regarded his present job as his least skilled. Two thought that their present jobs compared 'very well' to their least skilful job. It thus appears that whilst there are less skilled jobs in the service sector than those available at the Polytechnic, there are also more skilled jobs in the experience of these men.

Table 7.10
Best jobs in terms of skill.
(Those always in services)

Number	Industrial sector	Occupation	How present job compares?
1	Other services	Senior technician	Badly
2	Other services	School caretaker	Very badly
3	Other services	Assistant cook	Not much diff,
4	Military service	Officer	Can't compare

Note, the numbers are simply for reference in the text.

Table 7.11
Worst jobs in terms of skill.
(Those always in services)

Number	Industrial sector	Occupation	How present job compares?
1	Other services	Environmental officer	Very well
2	Present		
3	Other services	Hospital porter	Very well

Note, the numbers are simply for reference in the text.
One man refused to nominate a least skilful job.

The women with a domestic interlude work history pattern.

The women in this group spent a number of years in full-time work, partly in manufacturing, between leaving the education system and a 'domestic interlude'. One would expect this to be a period in which experience would be accumulated so that the most skilful job would be a full-time job held just before the temporary departure from the labour market. The sectoral location of this job is somewhat ambiguous as by this time it might be that women might be holding skilled jobs in the service sector, such as nurses or managing shops, as well as in the manufacturing sector as machinists for example. However, as most of these women are now doing unskilled work one would expect an unfavourable comparison between their most skilled job and their present one. Following on from this, the expectation regarding the least skilled job would be that it would be their present job or another part-time job held after the domestic interlude. Insofar as the job nominated as least skilled is not the present one would not expect a great deal of difference to be perceived between that job and the present.

Data pertaining to the job selected as most skilled is presented in table 7.12 It can be seen from this that the great majority of jobs selected as most skilful were full-time jobs, cases 1 to 18, and of these most were held when the woman was in her 20's, i.e. not immediately on entering the labour market but when the person had built up work skills.

Table 7.12:
Best job in terms of skill.
(Post-domestic women)

Number	Full-time?	Age when finished	Just before domestic break?	Sector	How present job compares?
1	Yes	23	Yes	Manufacturing	Badly
2	Yes	19	Yes	Manufacturing	Very badly
3	Yes	20	Yes	Manufacturing	Badly
4	Yes	20	Yes	Manufacturing	Badly
5	Yes	20	Yes	Distribution	Very badly
6	Yes	26	Yes	Distribution	Badly
7	Yes	22	Yes	Distribution	Very badly
8	Yes	19	Yes	Other services	Not much diff.
9	Yes	22	Yes	manufacturing	Don't know
10	Yes	25	Yes	Distribution	Very badly
11	Yes	19	Yes	Other services	Very badly
12	Yes	24	No	Commerce	Very badly
13	Yes	23	No	Distribution	Badly
14	Yes	32	No	Manufacturing	Badly
15	Yes	35	No	Manufacturing	Badly
16	Yes	47	No	Distribution	Badly
17	Yes	42	No	manufacturing	Badly
18	Yes	31	No	Polytechnic	*Present*
19	No	26	No	Distribution	Very badly
20	No	38	No	Manufacturing	Very badly
21	No	48	No	Other services	Very badly

Note, the numbers are simply for reference in the text.

In cases 1 to 11 these were jobs held before the 'domestic interlude'. Although there is not a perfect match here the general trend is clear. On the question of the sectoral location of these jobs there is a balance of service and manufacturing jobs, but only one woman chose her present job as the most skilled. As expected a fair degree of difference is perceived between the job nominated as most skilled and the current job, eight people thought their present job compared 'very badly' in skill terms to the selected job, ten thought 'badly' and only one saw 'not much difference'. (One did not know).

The data for the jobs selected as worst in skill terms is displayed in table 7.13. Here there are only twenty cases as one person claimed that she had never had a least skilful job.

Of the twenty people included here, eleven considered their present job to be their least skilful. A further six nominated a part-time job in services held after their domestic interlude as their least skilled. Only three chose a full-time job held in their youth. In two of these three cases the women had left her parental home whilst very young to take menial domestic work, the two jobs identified are as a nanny in a private household and a hotel chambermaid. The nine people who identified a job other than their present one as their least skilled split equally in their estimations of the comparative skill levels of the least skilled and the present job. Three felt the present job compared 'very well', three thought 'well', and three that there was 'not much difference'. The Polytechnic thus compares

Table 7.13
Worst jobs in terms of skill.
(Post domestic women)

Number	Present job?	P-T?	Age when finished	Sector	How present job compares?
1	Yes				
2	Yes				
3	Yes				
4	Yes				
5	Yes				
6	Yes				
7	Yes				
8	Yes				
9	Yes				
10	Yes				
11	Yes				
12	No	Yes	42	Distribution	Well
13	No	Yes	35	Other serv.	Not much diff.
14	No	Yes	49	Other serv.	Well
15	No	Yes	28	Distribution	Not much diff.
16	No	Yes	21	Distribution	Very well
17	No	Yes	26	Commerce	Not much diff.
18	No	No	16	Other serv.	Very well
19	No	No	*	Other serv.	Well
20	No	No	14	Manufacturing	Very well

Note, the numbers are simply for reference in the text.
*One woman's exact age is missing but she was in her teens.

rather more favourably than one might have expected.

Women with a non-domestic interlude work history pattern.

Again the question here is whether there is a significant difference between the pattern outlined in the preceding section and that of the two women who have worked continuously in paid employment? Have these two women been able to build a career in their unbroken working lives which would be reflected in their selection of the present job as their most skilful, with an earlier job, in which skills were being learnt, as their least skilful? Their work histories do not bear this out. Both chose earlier jobs as their most skilful, full-time jobs held when they were of mature years. What is ironic is that in both cases these jobs were left for 'domestic' reasons, one because of the woman's marriage, one because the woman had to care for her sick sister. In a sense then they suffered the same consequences as if they had taken a 'domestic interlude'.

In terms of jobs selected as least skilful, one woman chose her present job, which she had moved to after vacating her most skilful job because of a marriage rule. The other selected a job in her youth, like the two mentioned in the preceding section she also entered (paid) domestic work when very young as a general 'skivvy'.

So the simple absence of a 'domestic interlude' has not freed these two women from other constraints on womens' participation in the labour market as far as developing their skills is concerned.

Younger workers.

Whilst the common feature of the younger workers' labour market experience has been the instability of employment, in terms of skill their present occupations cannot be ignored. They are a mix of technicians, presumably exercising a degree of skill in specialist areas such as design or in workshops, and ancillary workers doing cleaning work or distributing food and cleaning in kitchens and eating places. From this one would expect the technicians to rate their present jobs highly in skill terms especially by contrast to the temporary jobs they have held in the past. But the relative skill level of the ancillary workers' present jobs compared to others they have had is rather open. On the one hand one would not expect them to identify their present jobs as the least skilful as many have held very poor jobs indeed. But on the other hand whether their present jobs are their *most* skilful is difficult to predict. However, insofar as another job is nominated as best for skill, one would not expect a great deal of difference between that job and the present.

Table 7.14 sets out the data for jobs selected as the most skilled. As expected the technicians here in the main identified their present job as their most skilful, with one exception. None of the people doing ancillary jobs made this nomination. Two points are of note. First, the jobs selected as most skilled do contrast to a degree with the people's present rather mundane jobs. The clearest case is the security man who had been a highly trained radio operator in the army, a caterer had held a supervisory position in a restaurant which had entailed considerable responsibility for staff and finance. Three others identified periods on training schemes where, by contrast to their present jobs (and indeed to school!) they felt to be learning something. This should not be overemphasised, the degree of difference between these jobs and the current one was, on the whole, not very great.

Table 7.14
Best jobs in terms of skill.
(Younger workers)

Number	Present job most skilled?	Current occupation	Most skilled job	How present job compares?
1	Yes	Academic tech.		
2	Yes	Academic tech.		
3	Yes	Academic tech.		
4	Yes	Academic tech.		
5	Yes	Academic tech.		
6	Yes	Mainten. tech.		
7	No	Academic tech.	Planning tech. local government	Not much diff.
8	No	Caterer	Catering supervisor	Not much diff.
9	No	Caterer	Training scheme Home for mentally handicapped	Not much diff.
10	No	Cleaner	Training scheme	Not much diff.
11	No	Attendant	Training scheme Painter	Badly
12	No	Security	Army, radio operator	Very badly

Note, the numbers are simply for reference in the text.

Table 7.15
Worst jobs in terms of skill.
(Younger workers)

Number	Present job least skilled?	Current occupation	Least skilled job	How present job compares?
1	Never	Academic tech.		
2	Never	Mainten. tech.		
3	Yes	Attendant		
4	No	Academic tech.	Sandwich course placement	Not much diff.
5	No	Academic tech.	Sandwich course placement	Very well
6	No	Academic tech.	Temporary barmaid	Very well
7	No	Academic tech.	Temporary shop assistant	Very well
8	No	Academic tech.	Assembly work Cowboy firm	Very well
9	No	Caterer	Caterer	Not much diff.
10	No	Caterer	Caterer	Very well
11	No	Cleaner	Assembly work Cowboy firm	Very well
12	No	Security	Storeman	Well

Note, the numbers are simply for reference in the text.

Jobs selected as least skilled are presented in table 7.15. Two people here claimed that they had never had a 'least skilled' job. It is of note that these people had had perhaps the least unstable work history of the younger workers. Neither of them had been unemployed, one had never been made redundant whilst although the other had been made redundant twice he had been fortunate to move straight into another job. Of the rest only one selected the present job as the least skilful, the rest nominating jobs they had held in the process of 'edging in' to the labour market.[7] Note that by comparison with these jobs, the present one compares favourably in many cases.

Other.

The one other case nominated her current full-time job as her most skilful and her part-time catering job at the Polytechnic as her least skilful.

The sample's selection of worst jobs for skill suggested that there might be a contrast between work in the secondary and service sectors but this was not supported by the selection of best jobs for skill. Again. although there was a degree of association between the selections and the individual's present occupation, this was accompanied by sufficient variation to suggest that past experience as well as the character of the current work task should be considered. The scheme of patterns of labour market participation has been successfully employed to throw light on the relations between such past experience and the sample's job estimations and comparisons in skill terms.

Notes.

1. Eight people took this view, six men and two women, six of them were academic technicians, one a maintenance technician and one a woman cleaner.
2. In the 1951 Census 60.6% of Coventry C.B.'s adult male population were classified as Social Class 3, Skilled Workers, in England and Wales the figure was 53%. In 1961 41.5% of Coventry C.B.'s adult males were allocated to the Socio-Economic Group Skilled Manual Workers, in England and Wales the figure was 31.6%. The corresponding figures for 1971 and 1981 were: Coventry 36.2% and 30.4% Skilled Manual Workers, England and Wales 29.8% and 26.3% (G.B. 1981). Figures taken from Procter, I. 1984a and 1984c.
3. Compare this with the findings on the occupational background of 'affluent' car and other workers in the Luton study in which the authors found that only 19% of semi-skilled car assemblers had been confined to semi-skilled work throughout their working lives. The authors argue that such workers chose to sacrifice skill and interest for the high wages of the car plant. (Goldthorpe, J. *et al* 1968: Pp.32).
4. This does not of course necessarily mean *unskilled*.
5. One did not feel able to make the comparison, perhaps because there was no comparison between his present and past jobs, but this is speculative.

6. One case is unfortunately missing on this item.
7. Cases 9 and 10 require a little more detail. Their present catering jobs are as cashiers, a step up the ladder from the washing up and cleaning tables positions they regarded as their least skilled.

8 Best and worst jobs in autonomy terms

The following questions were employed to explore the question of autonomy within a job.

Of the jobs you have had, which one most allowed you to get on with your work in your own way?

In this respect how does your present work compare with that job? Does it compare VERY BADLY/BADLY or is it NOT MUCH DIFFERENT?

Of the jobs you have had which one least allowed you to get on with your own work in your own way?

In this respect how does your present work compare with that job? Does it compare VERY WELL/WELL or is it NOT MUCH DIFFERENT?

The formulation of 'getting on with your own work in your own way' worked very well here. It did contain the presupposition that this was a virtue, an assumption which held in the vast majority of cases. There was one instance in which a different sense was put on the question when a respondent had job autonomy in abundance but resented this in the job in question because it implied to her lack of leadership from her superiors. This rather coloured her sense of the comparison with the present job.

Throughout this discussion the possible contrast between the factory context of manufacturing and current work in a service industry has been a continuing theme. The issue of job autonomy raises this question in a pointed way. Clearly work in factories covers much more than the stereotype of production line jobs but it is still associated with the pressures of production for profit, strong managerial control of the work process and the imperatives of mechanised production. The question is whether members of the sample with manufacturing experience perceived the factory in these terms and drew a contrast between this and the Polytechnic in which work provides a service rather than a tangible product, is non-mechanised and many jobs require individuals to work on there own with very distant management? This will be the central question in this section.

Tables 8.1 and 8.2 show the sectoral distributions of the jobs selected as best and worst for job autonomy. It will be noted that the total in table 8.2 excludes five people. This is because three felt they had never had a worst job in this respect,[1] one male cleaner did not know which job was worst in this respect, and one response was rather idiosyncratic (see the observation above).

Table 8.1
Best jobs for job autonomy
by industrial sector.

Sector	Number	Percent
Manufacturing	11	16.4
Construction	1	1.5
Distribution	7	10.4
Other services	3	4.5
Polytechnic	45	67.2
Total	67	100.0

Table 8.2
Worst jobs for job autonomy
by industrial sector.

Sector	Number	Percent
Extractive	1	1.6
Utilities	1	1.6
Manufacturing	28	45.2
Construction	4	6.5
Distribution	7	11.3
Transport	2	3.2
Other services	9	14.5
Polytechnic	6	9.7
Military service	2	3.2
Other unpaid	1	1.6
Casual worker	1	1.6
Total	62	100.0

This general sectoral distribution would suggest that the sample did perceive a contrast between the factory context of manufacturing and the Polytechnic with respect to job autonomy. What is immediately striking from table 8.1 is the large proportion of people who identified a Polytechnic job as their best for job autonomy and the small proportion who selected a manufacturing job. On the other hand, in table 8.2 we see that 45% of the sample chose a manufacturing job as their worst for autonomy but only 10% nominated a Polytechnic job.

The Polytechnic continues to be highly regarded in terms of job autonomy when we turn to the selection of the present job as best or worst in these terms. Only three people nominated their present job as their least autonomous job.[2] These were one female caterer and two security men. On the other hand forty-one people chose their present job as the most autonomous.[3] These included twenty-three of the thirty-three women and eighteen of the thirty-four men, split between the current occupations they hold as follows:

Table 8.3
Selection of present job as most autonomous.
By current occupational group.

Occupation	Yes	No
Academic tech.	14	4
Maintenance tech.	3	2
Protection	3	7
Cleaners	14	4
Caterers	4	8
Attendants	3	0
Total	41	25

The present job gives considerable autonomy to the technicians, cleaners and porters, although there are exceptions to each. On the other hand only a minority of caterers and security staff selected their present job as the most autonomous.

Although these initial items of information suggest the contrast of the factory and work at the Polytechnic, when we turn to analysis of this data in terms of the different patterns of work history we will see that the situation is not quite so straightforward as this.

The career development work history pattern.

It will be remembered that these eleven men had spent the bulk of their working lives prior to transferring to the Polytechnic in manufacturing industry. We would not expect them to identify such manufacturing jobs as their most autonomous and this is indeed the case. Eight of them selected their present job as their best for job autonomy. Of the remaining three, one selected a job working as an installation electrician for a firm of electrical subcontractors and one his job as a self-employed plumber. Only one defined a 'factory' job as his most autonomous, working as a laboratory technician in the aircraft industry.

However, the picture is not so straightforward when we turn to the jobs designated as involving least autonomy. On the one hand, ten of the eleven selected a job in manufacturing or construction as their least autonomous but, of these, eight chose the job in which they served their apprenticeship or its wartime equivalent. In drawing the contrast between their worst job for autonomy and their present job, which for many was a marked one,[4] it was clearly the contrast between their present autonomy and their status as apprentices which they had in mind. This thus adds a rather different aspect to the simple polarity of the factory and the Polytechnic.

The employment crisis work history pattern.

As tables 8.4 and 8.5 show there is considerable variety in the jobs selected by this group as those with most and least job autonomy. This is itself of some interest as one might have expected the job at the Polytechnic to compare well with the factory working environments in which these men have spent most of their working lives on the particular dimension of being able to control one's own work. However, no such simple dichotomy emerges.

Table 8.4
Best jobs for autonomy.
(Post-employment crisis).

Number	Best job	Present job	How present job compares
1	Present	Attendant	
2	Present	Cleaner	
3	Present	Attendant	
4	Cleaner at Poly.	Cleaner	Not much diff.
5	Stores foreman Massey-Ferguson	Protection	Badly
6	Assembly work Small engineering firm	Protection	Not much diff.
7	Fireman BL Courthouse Green	Protection	Not much diff.
8	Electrician BL Canley	Cleaner	Very badly
9	Engine fitter BL Canley	Protection	Very badly
10	Jig & tool fitter BL Canley	Cleaner	Not much diff.
11	Self-employed greengrocer	Protection	Very badly
12	Pre-apprenticeship	Protection	Not much diff.

Note, the numbers are simply for reference in the text.

Three people identified their present jobs as their most autonomous, one other selecting a job at the Polytechnic which he had held previous to his present job. On the other hand six people selected a job they had held in manufacturing industry. For them it appears that the move to the context of the Polytechnic from the factory has not entailed increasing job autonomy. Two men selected other jobs, one when he was self-employed and the other in a job when he was waiting to take up his apprenticeship and allowed considerable laxity whilst the time went by.

Whilst few people selected their present jobs as their most autonomous it is of note that the degree of difference between the current job and the nominated job is on the whole rather low. Three people said that their present job compared 'very badly' in terms of job autonomy, one 'badly' and five thought there was 'not much difference'.

Table 8.5
Worst jobs for autonomy.
(Post-employment crisis).

Number	Worst job	Present job	How present job compares
1	Present	Protection	
2	Present	Protection	
3	Protection Polytechnic	Protection	Very well
4	Chemical etcher Dunlop	Cleaner	Very well
5	Toolmaker Dunlop	Cleaner	Well
6	Track worker Fisher Ludlow	Protection	Very well
7	Warehouseman In transfer process	Attendant	Very well
8	Trainee/teaboy Engineering workshop	Attendant	Very well
9	Apprentice typesetter	Protection	Very well
10	Apprentice electrician	Cleaner	Don't know
11	Military service	Protection	Very well
12	Don't know		

Note, the numbers are simply for reference in the text.

Table 8.5 reveals a very varied picture. Cases 1 to 3 select jobs at the Polytechnic, 4 to 6 jobs in manufacturing, 7 a job held in the transition from manufacturing to services, 8 to 10 jobs held in the man's youth and case 11 selected his wartime army service. In cases where a job other than the present one was selected as least autonomous the current job compares favourably. One man said the present job compared 'well' and seven thought 'very well'. To the latter the 'don't know' can in fact be added as the man meant there was 'no comparison' between the two jobs.

It is perhaps here that one would have expected a strong contrast between autonomy at work in the secondary and service sectors. These were men who had spent many years in manufacturing industry, often in semi-skilled assembly or machining jobs. Their present jobs, although not particularly highly paid or skilled in terms of technical expertise or manual dexterity were not closely supervised in the sense of either mechanisation or an ever-present foreman. Yet only a minority of four selected a Polytechnic job as their most autonomous and three felt it was their least autonomous. On the other hand, six chose a factory production job as the most autonomous, whilst only three nominated such a job as the least autonomous. Whilst other factors enter in; self-employment, youth, military service and the trauma of transfer, the most important conclusion concerning these twelve men is the lack of a clear cut contrast. In a moment we will note a significant difference between these men and the next group of 'formerly sectorally mobile' male workers.

The formerly mobile work history pattern.

It will be remembered that these men had a varied work history which had taken them through a range of industries and jobs. The jobs they selected as best and worst for autonomy are presented in tables 8.6 and 8.7.

Table 8.6
Best jobs for autonomy.
(Formerly mobile men).

Number	Best job	Present job	How present job compares
1	Present	Protection	-
2	Cook Local government	Cleaner	Not much diff.
3	Self-employed cardboard reclaimer	Protection	Badly
4	Laboratory tech. Higher education	Academic tech.	Badly

Note, the numbers are simply for reference in the text.

Table 8.7
Worst jobs for autonomy.
(Formerly mobile men)

Number	Worst job	Present job	How present job compares
1	Salvage man Aircraft manufacture	Protection	Not much diff.
2	Coal miner	Cleaner	Very well
3	Machinist Electrical engineering	Protection	Very well
4	Electrician, optical lens manufacture	Academic tech.	Very well

Note, the numbers are simply for reference in the text.

Here there is a marked contrast between jobs selected as most and least autonomous. The former all lie in the service sector and whilst only one is located at the Polytechnic, the degree of difference between the selected job and the present job is not very marked. On the other hand, jobs nominated as least autonomous are in the secondary sector[5] and the comparison with the present job is marked in three of the four cases.

The difference in work history pattern between these men and the adjusters to employment crisis lies in the stability of past employment. This group stayed in jobs for a relatively short period of time whilst the former group were employed for long stretches of time at major manufacturing plants. As we have seen, the long term 'stayers' did not find the factory context particularly restrictive in terms of job autonomy, the formerly mobile men did.

The always services work history pattern.

As tables 8.8 and 8.9 show these four men rated their present jobs very highly in autonomy terms compared to other jobs they have held, mainly in the service sector.

Table 8.8
Best jobs for autonomy.
(Those always in services).

Number	Best job	Present job	How present job compares
1	Present	Academic tech.	
2	Present	Protection	
3	Present	Cleaner	
4	Laboratory tech. University	Academic tech.	Not much diff.

Note, the numbers are simply for reference in the text.

Table 8.9
Worst jobs for autonomy.
(Those always in services).

Number	Worst job	Present job	How present job compares
1	Military service	Academic tech.	Very well
2	Assistant cook	Protection	Very well
3	School caretaker	Cleaner	Missing
4	Local government officer	Academic tech.	Very well

Note, the numbers are simply for reference in the text.

Three of the four selected their present job as their most autonomous, with the exception seeing 'not much difference' between the job he selected and his current one. On the other hand for each of the jobs on which we have data for least autonomy the present job compared 'very well'.

The women with a domestic interlude work history pattern.

Table 8.10 sets out the data for jobs selected as giving most autonomy by women whose work history included a domestic interlude.

Table 8.10
Best jobs for autonomy.
(Post-domestic women).

Number	Best job	Present job	FT/ PT	Age when left	Pre/post domestic break	How present job compares
1	Present	Academic tech.				
2	Present	Caterer				
3	Present	Cleaner				
4	Present	Cleaner				
5	Present	Cleaner				
6	Present	Cleaner				
7	Present	Cleaner				
8	Present	Cleaner				
9	Present	Cleaner				
10	Present	Cleaner				
11	Present	Cleaner				
12	Present	Cleaner				
13	Present	Cleaner				
14	Secretary GEC	Caterer	FT	21	Pre	Very badly
15	Telephonist Massey-Ferguson	Caterer	FT	21	Pre	Not much diff.
16	Buyer Coop	Caterer	FT	25	Pre	Not much diff.
17	Chambermaid Hotel	Caterer	FT	*	Pre	Not much diff.
18	Hairdresser	Caterer	FT	19	Pre	Not much diff.
19	Shop assistant	Caterer	PT	22	Pre	Badly
20	Vending machine operator, machine tool manufacture	Cleaner	FT	45	Post	Not much diff.
21	Caterer College of Higher Education	Caterer	FT	35	Post	Not much diff.

Note the numbers are simply for reference in the text.
* Exact age not available, during her teens.

So thirteen of the twenty-one women here selected their present job as the one with most autonomy. These were overwhelmingly cleaners, ten of the eleven included here. It is noticeable that only one of the caterers nominated her present job as her most autonomous, seven did not. However, of the eight people who chose a job other than their current one, six said there was 'not much difference' between the job selected and their present employment. So in this respect the comparison is not a sharp one. Seven of the eight people selected a full-time job as their most autonomous, six of the eight a job prior to their domestic interlude, and five of the eight in service sector employment.

So, cleaning work at the Polytechnic clearly allows these women considerable job autonomy. The caterers, on the other hand, select predominantly full-time jobs they have held in the past, generally before their domestic interlude but spread between the secondary and service sectors. However the difference between these jobs and the present job is not a marked one.

Data relevant to jobs selected as worst in autonomy terms is listed in table 8.11. Three points become clear from this data. First, none of the women nominated their present job as their least autonomous. Second the present job compares very favourably with the job selected as least autonomous. Of the twenty who made a judgement, fifteen thought their present job compared 'very well' to their selected job, one thought 'well' and four saw 'not much difference'. Third, in fifteen of the nineteen cases in which the distinction applies,[6] the job selected was full-time. However, beyond this the picture becomes very mixed as the following figure indicates:

Full-time (15)	Pre-domestic (9)	Manufacturing (5)
		Services (4)
	Post-domestic (6)	Manufacturing (1)
		Services (5)
Part-time (4)	Pre-domestic (0)	
	Post-domestic (4)	Manufacturing (3)
		Services (1)

The jobs selected fell about equally between pre- and post-domestic break and between manufacturing and service sectors. So in terms of job autonomy these dimensions do not appear to be particularly significant.

The job autonomy offered by jobs at the Polytechnic stands out here. From the point of view of most autonomy, women doing cleaning work selected their present jobs as the most autonomous whilst the caterers saw little difference between their current job and the most autonomous. From the opposite point of view, none of the women saw her present job as her least autonomous but a great many of them drew a sharp contrast between their current work and their least autonomous job. Beyond this there seems little to distinguish between the jobs selected. Whilst the non-Polytechnic jobs chosen as most autonomous were largely full-time so were the great majority of least autonomous jobs. Again, there is no strong relationship with either the woman's domestic status or the industrial sector in which jobs were located. Both most autonomous and least autonomous jobs were spread both before and after the domestic interlude and across manufacturing and service industries.

Table 8.11
Worst job for autonomy.
(Post-domestic women).

Number	FT/ PT	Pre/post domestic break	Worst job	Present job	How present job compares?
1	FT	Pre	Research tech. GEC	Academic tech.	Not much diff.
2	FT	Pre	Machinist Clothing manuf.	Caterer	Very well
3	FT	Pre	Assembler Engineering	Caterer	Very well
4	FT	Pre	Product inspector GEC	Cleaner	Very well
5	FT	Pre	Assembler Aircraft manuf.	Cleaner	Very well
6	FT	Pre	Secretary Metal merchant	Cleaner	Very well
7	FT	Pre	Telephonist GPO	Caterer	Very well
8	FT	Pre	Nursing assistant Local government	Caterer	Not much diff.
9	FT	Pre	Shop assistant	Caterer	Not much diff.
10	FT	Post	Riveter Engineering	Cleaner	Very well
11	FT	Post	Childminder At home	Cleaner	Very well
12	FT	Post	Cleaner Contract cleaner	Cleaner	Very well
13	FT	Post	Waitress Restaurant	Cleaner	Very well
14	FT	Post	Telephonist GPO	Caterer	Very well
15	FT	Post	Caterer Restaurant	Caterer	Well
16	PT	Post	Caterer Glass manuf.	Cleaner	Very well
17	PT	Post	Solderer/wirer GEC	Caterer	Not much diff.
18	PT	Post	Solderer/wirer GEC	Cleaner	Very well
19	PT	Post	Caterer Polytechnic	Caterer	Very well
20	n/a	Post	Casual work	Cleaner	Very well

Note, the numbers are simply for reference in the text.
Note, one person refused to nominate a job with least autonomy.

Women with a non-domestic interlude work history pattern.

Both of these women identified their present jobs as the ones in which they had had most job autonomy. The jobs they selected as having least autonomy were a) a full-time job as a local government planning technician, vacated at 29, (comparison, present job was 'not much different' in autonomy terms) and b) a full-time job as a hospital nurse, vacated at 55, (comparison, present job compared 'very well' in autonomy terms). So there is little difference to the general pattern for older women noted above.

Younger workers.

Of the twelve younger workers, ten selected their present job as their best in terms of job autonomy. The two exceptions were an academic technician who nominated a previous post as a technician at the Polytechnic as her best job in this respect but felt there was 'not much difference' between the two and a caterer who chose her time on a youth training scheme in a home for the mentally handicapped as her most autonomous job. Her present job compared 'very badly' to this, but she is very much the exception here, also identifying her present job as her least autonomous, the only one to do so as is shown by table 8.12.

Table 8.12
Worst job for autonomy.
(Younger workers).

Number	Worst job	Present job	How present job compares
1	Present	Caterer	
2	Caterer	Caterer	Not much diff.
	Local government		
3	Warehouseman	Protection	Very well
	Parcels distribution		
4	Laboratory tech.	Academic tech.	Not much diff.
	Polytechnic		
5	Shop assistant	Academic tech.	Very well
6	Training course	Attendant	Very well
	Construction		
7	Laboratory tech.	Academic tech.	Very well
	Water supply		
8	Industrial artist	Academic tech.	Very well
	Electronics		
9	Assembly work	Academic tech.	Very well
	Photographic reproduction		
	carpenter		
	Motor components		
11	Government training scheme	Cleaner	Very well

Note the numbers are simply for reference in the text.
One person refused to nominate a least autonomous job.

One woman felt she could not select a least autonomous job as all her jobs as planning technician had been so similar. As mentioned, of the remaining eleven only one chose her present job, as a caterer. The striking feature of the rest is the degree of difference between the job selected as least autonomous and the present job. Eight of the ten said that the current job compared 'very well' in terms of autonomy, only two thought there was 'not much difference', these two selecting jobs involving the same kind of work they are still doing. The rest generally selected different kinds of work to their present positions, spread rather evenly between the secondary and service sectors.

Other.

The one other selected her present full-time job as her most autonomous and felt that she had never had a least autonomous job.

It is in terms of job autonomy where one might expect a strong contrast between the technical and organisational exigencies of the manufacturing factory and the provision of services. This is confirmed by a simple comparison of the sectoral location of best and worst jobs for autonomy but when we turn to more detailed work history experience a number of complicating factors enter in. For example, although the career developers located their best jobs here in their present positions and their worst jobs in manufacturing, this was found to be more to do with the pattern of accumulating skills over time as their worst jobs, in this respect, tended to be those in which they were learning their trade. Again the men who had worked for many years in manufacturing did not draw a straightforward contrast. For many of them the factory had been a place of considerable work autonomy whilst the factors contributing to lack of autonomy were multiple. Clearly, the women cleaners in the sample found considerable autonomy from supervision in their present jobs. But insofar as the present job was not selected as most autonomous, the women workers nominated jobs in both manufacturing and services as both most and least autonomous. Here, the particular conditions of the Polytechnic seem especially important. Very strong comparisons were drawn between the present job and the least autonomous job, irrespective of whether the latter was located in the secondary or service sector. This also applied to the younger workers who again drew sharp comparisons between present employment and other jobs, the latter being found across industrial sectors. Thus, although the sectoral difference is a factor here, it should be placed in the context of the overall pattern of peoples' working lives.

Notes.

1. These were all women, an academic technician, a cleaner and a caterer.
2. This differs from the 6 in table 8.2 because some people nominated a job at the Polytechnic which they no longer hold.
3. This differs from the 45 in table 8.1 because some people nominated a job at the Polytechnic which they no longer hold.
4. Seven of the ten saw their present job as comparing 'very well' with their worst, two as 'well' and one saw them as incomparable.

5. If we can stretch this to include coal mining at this point.
6. The exceptions are the woman who felt she had never had a job without autonomy and the woman who selected her years working casually at a variety of jobs as her least autonomous.

9 Best and worst jobs for trade unionism

The previous three chapters have discussed work aspects applicable to all paid jobs. Trade union representation is rather different in that membership is not universal. In fact two sample members were not currently and never had been union members whilst a further seventeen had only been union members in one job. This chapter will begin with a brief analysis of these sample members who were not in a position to compare best and worst jobs for trade unionism.

For each job in which a person was not a union member the respondent was asked why this was the case. This was an open-ended question to which the responses were coded later according to the following scheme:

a) No union at the workplace.

b) Not asked to join. The presumption here is that there was a trade union at the workplace but this was not always clear to the respondents.

c) Union membership was not applicable in this job, according to the respondent, because it was a *temporary* job.

d) Union membership was not applicable in this job, according to the respondent, because it was *self-employment*.

e) Union membership was not applicable in this job, according to the respondent, because it was a *small firm*.

f) Union membership was not applicable in this job, according to the respondent, because it was *domestic employment*.

g) The respondent did not wish to join a trade union.

h) The respondent had resigned from a union.

The two people who had never been members of a trade union were both women academic technicians in their mid-twenties. Their reason for not joining a trade union in jobs held prior to their present one was in each case that the job was temporary in one way or another. This was the reason why one of them was not a union member in her present job, the other gave the fact that she had never been asked to join as her reason.

Seventeen people had been union members in only one job. In sixteen cases this job was the present job, the exception being a male academic technician who had once been a member of the National Association of Local Government Officers (NALGO) but had resigned because of its 'total inefficiency'. His overall work history was the always service industry pattern. Of the seventeen, fourteen were women and three men. Their current occupations were academic technicians (5), maintenance technicians (1), cleaners (6), caterers (4) and porters (1). Their patterns of labour market participation were career developers (1), post-domestic women (7), non-domestic women (1), younger workers (6), always services (1) and other (1). Table 9.1 displays the reasons why these people said they had not been union members in the paid jobs they had held.

Table 9.1
Reasons for union non-membership.
(Those in a union only once)

Reason	Number	Percent
No union	19	33.9
Not asked	20	35.7
Temporary	6	10.7
Self-employment	1	1.8
Small firm	6	10.7
Domestic	1	1.8
Did not want to join	1	1.8
Resigned	2	3.6
Total	56	100.0

Almost all the reasons offered related to the absence of the opportunity to join a union rather than a desire not to join, the latter applying in only three of the jobs covered here.[1]

So, the great majority of people who had only been union members in one job had been recruited as unionists in their present employment. They were predominantly women, with a work history pattern of either returning to employment after a domestic interlude or being younger workers. In the past they had not been union members because of the absence of unionism in the places they had worked.

We now turn to sample members who had been union members in more than one job. The following questions were employed to discuss this topic:

Looking at the jobs you have had, when do you think you were best served by a trade union?

Can you say, very briefly, what was good about the union at the time?

How does your present union compare with your most effective union in the past? Does it compare VERY BADLY/BADLY or is it NOT MUCH DIFFERENT?

Can you briefly say why your present union does not serve you as well as the one in the past?

Again, looking at the jobs you have had, when do you think you were least well served by a trade union?

Can you say, very briefly, why the union was not very good for you at that time?

How does your present union compare with your least effective union in the past? Does it compare VERY WELL/WELL or is it NOT MUCH DIFFERENT?

Can you briefly say why your present union serves you better than the one in the past?

The formulation of the questions here attempted to address the individual's sense of being served by a union, irrespective of what the person defined as the objectives of trade unionism. A relatively large number of people refused to reply to some or all of these questions, either on the grounds that they had never been well served by a union or that they had never been poorly served by a union. To some extent this had been anticipated, both in terms of past research and on the basis of piloting the questionnaire. Many people's attitudes to unionism, whilst being channelled by the mass media, are rooted in particular instances in which they have actually come into contact with 'the union'.[2] It was because of this that the above open-ended questions were included, to sketch the reasoning behind the response.

The questions on best and worst jobs for unionism evoked a high rate of non-response, fifteen people declined to answer one or both of these questions. These fell into two groups, those who refused to comment because of their hostility to unions and those who declined to nominate a worst union because of their support for unions. In the first group are five people who when asked when they were best served by a trade union replied with an emphatic 'never!' and then, although having plenty to say on the matter, would not nominate a worst trade union. To these can be added two people who gave the same response to the best trade union question but would nominate a worst union. In addition, one person answered the best union question but then went on to elaborate how he only paid lip service to trade unionism and how he believed unions to be the main reason for a host of economic problems. Thus eight people were generally hostile to trade unions. On the other hand seven people did not nominate a worst trade union because they said that they had always been well served by their unions. Of these, four replied to the question with an emphatic 'never', three with the rather less strong response of 'don't know'.

Information relating to these people is presented in tables 9.2 and 9.3. The picture which emerges here is unclear in terms of sex, occupation and work history. Of the eight people generally hostile to unions, six are men whilst of the seven generally favourable, four are women, so there is an inconsistent tendency for these general attitudes to vary by

Table 9.2
Non-responders to trade unionism questions.
(Those generally hostile to trade unions).

(a) Number	(b) Sex	(c) Current occupation	(d) Work history pattern	(e) Percent of years in union	(f) Union posts held	(g) Branch meeting attendance
1	Female	Caterer	Domestic inter	13		1,1
2	Male	Acad. tech.	Career dev	100	Shop steward	1,1,1
3	Male	Maint. tech	Career dev	70		5,1
4	Male	Porter	Employ crisis	97		4,1
5	Female	Caterer	Domestic inter	87		1,1,1
6	Male	Acad. tech	Career dev	49		5,5
7	Male	Security	All service	83		1,1,1,1
8	Male	Maint. tech.	Younger	100		1,1,1

Table 9.3
Non-responders to trade unionism questions.
(Those generally favourable to trade unions).

(a) Number	(b) Sex	(c) Current occupation	(d) Work history pattern	(e) Percent of years in union	(f) Union posts held	(g) Branch meeting attendance
9	Male	Acad. tech.	Career dev	100	Shop steward Convenor District cmttee	3,3,3,3,3
10	Female	Cleaner	Domestic inter	100		3,3,3,1,1
11	Female	Cleaner	Domestic inter	85	Shop steward	9,9,3
12	Male	Acad. tech.	Career dev	87	Shop steward Convenor District cmttee	3,3,3,3,1
13	Female	Cleaner	Domestic inter	50		1,1,1
14	Male	Cleaner	Employ crisis	89		1,1,1,1,1,1,1
15	Female	Cleaner	Domestic inter	42		1,1,1

Notes for tables 9.2 & 9.3:

1) The numbers are simply for reference in the text.
2) Column (e) shows the number of years as a trade union member as a percentage of the individual's total number of years in paid, civilian employment.
3) Column (f) shows the trade union positions held by the individual throughout working life.
4) Column (g) shows the frequency of attendance at branch meetings in the jobs in which the individual was a trade union member. The key is: 1 Never attended; 3 Attended almost every meeting; 4 Attended every 6 months or so; 5 Attended about once a year; 6 Attended less than once a year; 9 Missing data.

sex. Again, the pattern in terms of current occupation is rather ambiguous, those generally hostile are spread amongst the occupational groups with a hint that technicians might be prominent. But on the other hand two technicians appear in the generally favourable table. The work history patterns of the fifteen people are very varied.

Moving on to these people's trade union history, it is noticeable that the majority, both generally hostile and favourable, have been union members throughout most of their working lives. However, those generally hostile have not been active unionists, one of them had been a shop steward but the fact that he did not attend branch meetings suggests that he was not particularly active. On the other hand, four of those generally favourable have a record of union involvement. Two of them had held quite senior positions in the running of their unions and another is still an active shop steward. All four had at times in their careers regularly attended union meetings. These four were the people who positively refused to nominate a worst trade union, the other three people who were generally favourable did not know when they had been least well served by a union because they had never had any complaint. By contrast, of the eight people who were generally hostile to unions, five related a particular grievance to the interviewer. In two cases this referred to past experiences which had disillusioned them. One woman and her colleagues at the time had, in her view, been let down by a union official when they had protested at their transfer to another job, the official siding with management. One man who had worked for many years at the Canley plant complained of the strike activity he had been involved in there. This applied to another man who had worked at the Dunlop plant, who also felt that his present union (NALGO), was not organised to represent technical workers like himself. In this he was joined by two other people who felt that their present union was geared to defend the interests of occupations other than their own, in one case again a technician who felt that NALGO did not serve the interests of craftworkers, in the other case a protection officer who felt that the Transport and General Workers Union (T and G) was not interested in monthly paid staff.

Those who refused to nominate a best or worst trade union had generally had considerable experience of union membership. Those who were generally hostile to unions often expressed particular grievances arising out of this experience, those generally favourable had often been active in the union movement. But the direction of causation here is unclear, whether unfavourable attitudes lead to the expression of grievance or *vice versa*, or whether favourable attitudes lead to activism or *vice versa* cannot be determined from the data at hand.

The upshot of these points on the extent of membership and question refusal is that our data on trade unionism is more restricted in terms of the proportion of the sample making choices and judgements than for the other aspects of jobs we have considered. Clearly the two people who had never been union members and the seventeen who had been members in only one job were not in a position to compare different experiences of union membership. Added to these are the seven people who refused to nominate a best union and the thirteen who were unwilling to select a worst union. Thus forty-one members of the sample answered the best union question and thirty-five selected a worst union.

In Britain generally, union membership varies considerably between industries. Using data for 1979, R. Price and G.S. Bain (1983: Pp. 52) report that union density[3] in the public sector was 82.4%, amongst manual workers in manufacturing it was 80.3%, amongst white-collar workers in manufacturing it was 43.7%, whilst in private services it was only 16.7%. Thus unionism has its areas of strength in the public services and manufacturing and its area of weakness in private service industries. This provokes the question whether people feel to be best served by their union in its area of strength, worst served in the area of low union membership? A further possible contrast is between unionism in manufacturing and the public services. This might be particularly the case in Coventry. From 1939 until the mid-seventies Coventry's large manufacturing plants were centres of highly developed shop-floor union organisation and representation. They have been the subjects of a whole series of studies of modern trade unionism in action.[4] Although Coventry's public services are unionised, no author has argued that this is particularly remarkable as compared to the country at large. One wonders, then, whether the sample members, many of whom had formerly been employed in these bastions of trade unionism, drew a contrast between unionism in manufacturing and in the public services?

To explore these questions we can turn to the jobs selected as best and worst for unionism by the sample, presented by industrial sector in tables 9.4 and 9.5.

Table 9.4
Location of best jobs for trade unionism by industrial sector.

Sector	Number	Percent
Manufacturing	17	41.5
Construction	2	4.9
Other services	2	4.9
Polytechnic	19	46.3
Casual worker	1	1.4
Total	41	100

Table 9.5
Location of worst jobs for trade unionism by industrial sector.

Sector	Number	Percent
Extractive	1	2.9
Manufacturing	10	28.6
Construction	3	8.6
Distribution	2	5.7
Transport	1	2.9
Other services	6	17.1
Polytechnic	12	34.3
Total	35	100

Table 9.4 shows that no respondent chose private service industry as the location of the best job, these predominantly lying in manufacturing and the public sector of 'Other services' and the Polytechnic. However, when we turn to jobs nominated as the worst for unionism we also find that few of them are found in the private services, only three people making this choice. Once again it is manufacturing and public services which predominate. So there is no simple division between best jobs in areas of union strength and worst jobs in areas of weakness. Rather, almost all the jobs selected as best and worst

for unionism lay in the manufacturing and public service sectors. Neither do they split between these two sectors in terms of best in manufacturing and worst in public services or *vice versa*. Rather, 42% of best jobs were in manufacturing, 51% in public services. Worst jobs are a little more concentrated in public services (51%), but 29% were located in manufacturing.

A further feature of note here is that, within public services, the Polytechnic is predominant in the selection of both best and worst jobs. We can analyse this further by using tables 9.6 and 9.7 which document which people selected the present job as best and worst in terms of sex and current occupation. Once again a very mixed picture emerges. Seventeen people considered their present job as the one in which they have been best served by a union, split equally between men and women and distributed quite evenly between the various occupational groups. Eleven people selected their present job as their worst for unionism, four women and seven men, caterers and technicians being noticeable here although the numbers are small and in each case the number not choosing the present job is substantial.

Table 9.6
Present job chosen as best for trade unionism.
By sex and occupational group.

Sex/occupation	Yes	No	Total
All women	9	6	15
All men	8	18	26
Women acad. tech.	2	0	2
Women cleaners	4	2	6
Caterers	3	4	7
Men acad. tech.	1	7	8
Maint. tech.	0	2	2
Protection	4	5	9
Men cleaners	3	3	6
Attendants	0	1	1
Total	17	24	41

Table 9.7
Present job chosen as worst for trade unionism.
By sex and occupational group.

Sex/occupation	Yes	No	Total
All women	4	8	12
All men	7	16	23
Women acad. tech.	0	2	2
Women cleaners	0	3	3
Caterers	4	3	7
Men acad. tech.	3	2	5
Maint. tech.	0	2	2
Protection	3	7	10
Men cleaners	1	4	5
Attendants	0	1	1
Total	11	24	35

Having found little consistency in the sample members' responses to these questions in terms of differences between industrial sector or sex and current occupation, we can again turn to the relationship between judgements on trade unionism and the individual's work history.

The career development work history pattern.

This sub-group consists of eleven men, one of whom had been a union member in only one of his jobs (the current one), an electrician who had previously worked for small electrical contractors. Two people refused to answer either the best or worst trade union questions and a further three did not reply to the worst trade union question. Of the eight who nominated their best job for trade union representation, two selected a job at the Polytechnic (one his present job, one a previous Polytechnic job) whilst six selected a job in manufacturing or construction. Of the five who nominated their worst job for union representation, two selected their present job and three a former job in the secondary sector. Thus, on the whole, the effectiveness of trade union representation had not run in parallel with the development of these men's careers. The reason for this came out quite clearly in the responses to the open-ended questions. In shifting to service industry they had been able to transfer their skills but not the *trade union* which had earlier represented those skills. The feeling of most of these workers was that NALGO was indeed a union for local government officers, not technical and craft workers like themselves. Details of the cases are:

1) Refused to answer either question but stated that NALGO did not represent engineers.[5]

2) Refused to answer either question but stated that NALGO did not represent electricians. The ETU was 'his union'.

3) A former active trade unionist in the motor components industry, felt that NALGO 'detached' and could not represent his job.

4) A former active unionist in the aerospace industry who felt that NALGO was 'an old pals' act', i.e. again the sense of distance from the union though here not expressed in occupational terms.

5) A building craftworker who disliked collective grading systems because they did not distinguish or reward different degrees of skill.

6) A former active unionist who selected a print union as his most effective because of its specialist membership and representation, by contrast NALGO was 'Too large, all embracing...does not seem relevant'.

7) A former active trade unionist who contrasted his experience in establishing a draughtsman's union at GEC with the 'formal structure' of NALGO.

The theme here is the sense that the occupational interests of these workers are distant from the objectives and capacity of the union they are currently members of and which is recognised by the institution in which they work. This is linked to their work history pattern, they have carried over an identification with unions which represent the *occupations* in which they are trained and skilled and feel little common identification with the branch of *industry* (local government) by which their present union is defined. It is of note here that four of these seven have in the past been active in the union movement but are so no longer, whilst one other is a trade union representative at present but does not attend branch meetings of NALGO.

There are three exceptions to this pattern. One man was generally hostile to trade unions and this had little to do with any sense of disjuncture between occupation and current union representation. A second appreciated the regular annual pay rises that NALGO negotiated, a point mentioned by the third who identified his former craft union as his worst because it had resisted him leaving for NALGO when he came to the Polytechnic. But the latter was the only case in which a strong feeling of identification with the person's present union came across.

The employment crisis work history pattern.

This sub-group consists of twelve men, only one of whom refused to answer the trade union questions, and he expressed his views even if not in direct response to the questions. As we have noted, prior to redundancy and a process of transfer to their present jobs in service industry, these men had held long term, stable jobs in manufacturing. The plants in which they had worked had been highly unionised, the exemplars of the strong plant level union organisation of Coventry in the post-war period. In seven cases it was this job which was identified as the one where the person was best served by a trade union. The present job compared 'very badly' in these terms in four of these cases, 'badly' in one, 'not much different' in one case, and one man did not know. In one other case an earlier job in manufacturing was selected (present job compared 'badly'). However, the picture is not so simple as this image of movement from well organised manufacturing plants to a service industry with 'inferior' unionisation. For one thing, in three cases the present job was nominated as the one in which the individual was best represented by a trade union. Again, in three cases the job immediately prior to leaving manufacturing was selected as the worst for union representation. The person who refused to answer can also be added here as he was highly critical of union activity at the B.L. Canley plant. Three other men identified an earlier job in manufacturing as their worst for unionism. On the other hand only three identified their present job as the one in which they were least well served by a trade union and only one a job in the process of transfer to service sector employment.[6]

So a rather mixed picture appears. However, just as in the case of the career developers preoccupation with the mismatch between their occupational skills and the union which represented them, so here a common theme emerges from the reasons offered by these men in connection with their selections. This is their concern with the quality of shop-floor representation, the degree to which the individual feels to be in communication with and served by identifiable union personnel at the place of work. This works both ways, in some cases it is the basis of positively evaluating the union, in others the reason for complaint. Either way it is the criterion on which these men judged the adequacy of union representation. Brief sketches of the responses to these questions will serve to establish this theme.

1) This man selected his time at Massey-Fergerson as his best for trade unionism saying 'The shop stewards took an interest, informed you what was going on. They were good negotiators without strike action.' In the present job: 'No one seems to worry about the union here. Its almost non-existent, I just pay my subs.' He did not, however, see his present job as his worst for unionism, selecting a job as a long distance lorry driver because he was away from home and had no contact with his steward. He felt the present job marginally better than this.

2) This man selected his job at BL Courthouse Green as his best for trade unionism because of the redundancy terms negotiated. However he felt that the present job compared 'very badly', indeed was his worst for unionism, because 'They don't seem interested in us.'

3) The BL Canley plant was selected here as the best for unionism because 'When there was any bother on, I always had good service.' The present job is the worst for unionism because 'They have done nothing for me personally.' And he had needed assistance as his wages fell in a job reorganisation, a fact he felt most aggrieved about.

4) By contrast the next man nominated his present job as his best for unionism on the grounds that his income had increased substantially since he started in 1971, but, returning to the theme, his time at Dunlop Aviation in the 1960's was identified as his worst for unionism because he was badly let down by the convenor. In a dispute, the convenor had taken the management side and he was blamed for something that was not his fault.

5) The next case also identified his present job as the one in which he was best served by a trade union, commenting that the union provided security but also that the union representatives 'listened to the workforce rather than dictating to them' and giving the example of cooperation between management, union officials and workforce over a job reorganisation. (It is ironic that that this was the same event over which case 3 above felt so aggrieved). By contrast, his worst job for unionism was at the Rootes car plant when in his view the union negotiators felt that the security staff (of which he was one) was overpaid.

6) This man selected the BL Canley plant as his best job for unionism commenting that 'The convenor was responsible, interested in getting full-time employment for the men...very fair.' His worst job for unionism was in a temporary job with a parcel delivery firm in the process of transition to his present job. There the union representatives were not interested in temporary workers. He felt that things were only marginally better in his present job as there was 'still not much contact' and his present job compared 'very badly' with his Canley job because 'I rarely see the rep.'.

7) The BL Canley plant is again selected here on the grounds of the union's 'support' and 'the clever negotiators...the Communist shop stewards were the fairest in dealing with management and didn't try to double cross the blokes they were representing'. The present job compared 'very badly', an instance was quoted of a negotiator 'changing horses' midway through discussion with management. The present job is the worst for unionism because 'I don't get to see any representatives'.

8) In this case the reasons for nominating the best and worst jobs for unionism were rather tangential to the main theme here. A job at GEC in the early 50's was selected as the best because of the unions' district influence on the Coventry Toolroom Agreement[7] whilst the dying years of the Canley plant were chosen as the worst for unionism because the man disagreed with the overall union strategy of trying to protect an inefficient, loss-making plant. But when asked to compare his present job to these experiences he returned to the theme saying that 'they were out of touch' and instancing the union's lack of appreciation of a differentials grievance.

9) This man chose his early years at the Canley plant (commenced 1959) as his best for unionism 'because your grievances were sorted out'. He 'did not know' how his present job compared nor what his worst job for unionism had been, the reason being that he was preoccupied with telling the interviewer about his unhappiness with his present job. It appeared, however, that he had not considered approaching a union representative about his present situation.

10) The present job was selected as best because 'we are well represented as regards conditions, hours....'. His time as an apprentice coach builder was the job in which he was least well served by a union in that he had 'little contact' with the union.

In two cases this theme of the quality of shop-floor representation was marginal to the judgements made:

11) This man refused to make judgements about his best and worst unionism but had plenty to say about the money lost in strike activity at the Canley plant and how he only regarded his present union membership as an insurance in case of accident. He did however also comment that he 'never saw anyone'.

12) In this case the same union/job was selected as best and worst for unionism in that whilst in the printing trade the union provided a 'good service although I never personally used it' but at the time of redundancy the man felt the union could have 'done a better deal'.

In the majority of these cases the way in which shop floor representatives contact, inform, consult and act in the interests of the individual worker is the standard used by these men in estimating when they were well served or poorly served by a union and in comparing their present jobs to earlier ones. Whilst it is not the case that these men all made a nice neat contrast between their union representation in the strongholds of the Coventry labour movement where they had previously worked and their present service sector jobs, it is plausible to argue that this *criterion* is a product of their past employment experience in such contexts. Whether or not the particular worker felt that he was well served by strong shop floor organisation in the past or is at present, his expectation is that workers should be in close contact with known union personnel as part of the everyday routine of work. Whilst the data available here cannot demonstrate this, it is possible that such an expectation is rooted in the past experience of work in workplaces with well developed union organisation.

The formerly mobile work history pattern.

Although these four men had not worked consistently in the large scale manufacturing plants which characterised the industrial history of the last sub-group all of them had considerable experience of such plants and the well organised shop floor unionism found there. So we would not be surprised to find a similar pattern of response to the trade union questions. This is the case to a considerable degree:

1) This man nominated his present union as his best on the grounds that the union was 'very good when there's any trouble, they act quickly and listen'. His worst union was

when he worked as a milk roundsman, he now regrets that he did not take the problems he was having with his supposed colleagues to the union. The present union compares 'very well' to this because 'they are always on hand to help and advise'.

2) A second man selected his job as a machinist at English Electric as the one when he was best served by a union 'because I had lots of contact with the rep., he was on the same section, he was always available and always had time'. His present union was 'not much different' to this and compared 'very well' to his worst union. This was in his job as a hospital porter in which he mentioned (unspecified) 'problems within the union'. By contrast the present union held more meetings and 'the rep is always available'.

3) The third case is more ambiguous, he nominated his job as a maintenance electrician at a car components firm as his best for unionism on the rather obscure grounds that, although there were lots of strikes there he was not affected because in a different union, the ETU. His worst job for unionism was again as an electrician at Rugby Cement because the company did not recognise the ETU and thus he had no representative on the site, returning to the theme of shop floor representation.

4) In the case of the last man here we have very inadequate data because of interviewing difficulties. He had only been in a union in his present job and in short spells as a coal miner. He put the present union as marginally ahead of the NUM. However, as he had worked as a miner for only brief periods of time it is perhaps more accurate to say that he had only been in a union in the present job.

The always services work history pattern.

Here are a sub-group with a very different industrial history to the men we have so far considered in that their careers have been marked by the absence of experience of the highly unionised industrial plant. If the suggestion that this experience predisposes workers to have an expectation that the quality of shop floor representation is the measure of being served by a union is persuasive, then we would expect this to be absent in these cases. In three of the four cases this expectation holds:

1) In the first case the man nominated his time as a university laboratory technician as his best for unionism on the grounds that the union had successfully negotiated the removal of car parking fees for employees at the workplace. This compared 'very well' with NALGO (his worst union) from which he had recently resigned because of the union's financial contribution to the miners' strike.

2) The second man was of the view that he had never been well served by a trade union and put his present union (the T and G) as his worst because he was paid on a monthly basis and the T and G were geared to negotiate on issues concerning weekly paid staff.

3) The third man had only been a union member in one job, an earlier post at the Polytechnic. He had resigned because of the 'total inefficiency' of the union. We have no further details of what he meant by this.

4) The last case does return us to the theme of shop floor representation. This man selected his present union as both his best and worst. What he was doing was selecting particular instances in which he had had contact with the union. The union had given him assistance in a period of illness but had failed to act on a differentials grievance, in each case it was his immediate contact with the union which served as his measure of its worth rather than the more 'abstract' issues raised by the other three people.

The women with a domestic interlude work history pattern.

Of the twenty-one women here, seven had been union members in only one job and in all cases this was their present job. Of the remaining fourteen, two said that they had never been well served, one of these refusing to say any more, the other nominating her job as a components inspector at GEC as her worst for trade unionism on the grounds that the union was 'Not interested in the lower life of the factory floor'. From the context it was clear that by 'lower life' she meant women workers, a theme we will return to in a moment. In contrast to these two women who displayed generally hostile attitudes to trade unions were four women who 'did not know' or had never been least well served by a union. Their responses to this question were an emphatic 'never', 'never, I've always been fully satisfied', 'I don't know because I've never had reason to call on a union' and 'I don't know...I can't remember any time when I've been let down'. Two of these women nominated their present job as the one in which they were best served by a union. Their reasons were: 'I don't know, I don't have a lot to do with them, I'm just a member' and 'I don't have any reason to call on a union but if I need them they are there'. The two who nominated a previous job as their best for unionism said that their present job was 'not much different' to the one they chose. They selected the earlier jobs because 'They got us good rises and good working conditions', and the union 'looked after me when I had an accident'. What is striking about these responses is that, with the exception of the woman who had *never* had a worst union, cited the accident support as her instance of union service and who was in fact an active shop steward, they have a certain blandness about them. Whilst these women were diffusely positive in their support for unionism this rested on the absence of reason for complaint rather than definite reasons for support. This point is worthy of note in contrast to the responses of the remaining eight women who nominated both a job in which they were best served and one in which they were least well served by a union. As the following case notes display, what is noticeable here is the strong feelings the questions evoked about specific issues which often related to conditions particular to women's employment. This led the women to draw strong contrasts between selected jobs and the present job in comparisons in both best and worst directions.

Four women selected a job prior to their present one as their best for unionism. In each case this was accompanied by the present job being nominated as the worst for unionism.

1) This woman chose her job as a wirer at GEC as her best union because 'if there was anything wrong on the job the shop steward would sort it out'. Her present union compared 'very badly' to this job, she commented: 'You can't get hold of a union woman in this job and when you do it doesn't come to anything. They are more with management's point of view than the workers' '. There is no reference to issues specific to women's employment here, in this she is the exception.

2) In this case a job as a caterer at a technical college was selected as the best for unionism because just as she took the job a dispute broke out and even though she was new she was paid strike pay. What impressed her was 'We were all treated the same'. Her present job compared 'badly' to this job, the union 'don't want to hear our point of view'. What she had specifically in mind was the point of view of workers who are laid off during the vacation and yet still have to continue paying union subscriptions. She had been to the T and G office about this but 'they don't want to know'.

3) The third case could give no definite reason why she selected a job as a local authority nursery assistant but held strong views on why her present job was worst for unionism: 'They don't take the interests of part-time workers seriously', her example being the payment of full subscriptions from a part-time wage. She believed that subscriptions should be proportional to hours worked.

4) This woman selected her job at GEC as her best for unionism because of the unions success in getting better wages and conditions. Her present job compared 'very badly', partly because of poor shop floor representation (she said she had seen her shop steward once in nine years) but also because the union was 'not interested' in part-time workers.

Finally, four workers selected their present job as their best for unionism, drawing strong contrasts to earlier experiences which again often relate to the conditions of women's employment.

5) This woman's initial response to the best union question was 'They have all been useless to me' but with a prompt she selected her present union because on the one occasion she had needed help she had obtained it. The reason for her first reaction quickly became clear when she was asked for her worst union. In this she had no doubt, a job as a telephonist at Dunlop in her teens. She and her colleagues had a grievance which she 'naively' took to the union. She was 'manipulated' by the official who 'acted dishonestly' and left her 'high and dry' and she had to resign the job. Although this woman did not go into the details of this incident the interviewer had the distinct impression that it was because she had been a young woman that the official had treated her so cavalierly.

6) Again the reason why the next woman selected her present job as her best were very much connected to her earlier experience as a needle stringer at a firm manufacturing knitting machines. She worked on the evening shift where the workers found that the day shift constantly left the 'rotten work' for her and her fellow part-timers. The union failed to take action on this, because, in her view, they were part-timers. Furthermore when the firm sought redundancies it was the evening shift that went, without union objection.[8]

7) In this case the woman selected her present job because the union had done something for part-time workers, getting them a retainer fee for the summer vacation. She contrasted this to her time as an apprentice hairdresser when her boss would not permit union membership whilst the working conditions were 'dreadful'.

8) The last case does not involve a particular instance or a issue particular to women. She selected assembly work at GEC as her worst job for unionism because 'they never seemed to tell us anything, we didn't even realise who the union reps were'. Her present job compared 'very well' to this because 'we know who the rep is and we get information'. She selected the present job as her best for unionism because of the wage rises the union negotiated.

As in the cases of the male workers we have already discussed there is no simple pattern of response amongst these women workers. Some define the present job as their best for unionism, others the worst. Again this is repeated for similar earlier jobs, of the four who instanced jobs at GEC,[9] two gave this as their best job, two as their worst! Clearly the experience of unionism can vary enormously between individuals at the same workplace or doing the same job. Yet amongst these women workers who were asked to compare unionism in different jobs a theme emerges, a strength of feeling linked to particular experiences (both positive and negative), often connected to their status as *women* workers. Whilst a minority were either diffusely hostile or supportive to trade unions the majority drew strong contrasts between their union experiences, usually relating these judgements to issues which had had a considerable personal impact upon them.

Women with a non-domestic interlude work history pattern.

Once again there seems to be little to differentiate these two women from the rest, one had first joined a trade union on coming to the Polytechnic; the other neatly illustrates the theme above about women being concerned with specific issues which had influenced their working lives as women. She nominated a job as a local authority planning technician as her worst for unionism for the apparently rather obscure reason that, as the employer recognised her capacity to do what was usually regarded as a 'man's' job, she did not need a union. This becomes clear in her comparison with her present job. In her view her present employer does not respect her competence but she finds her union supportive and selects it as her best union because of 'The fact that it looks upon the sexes as equal'.

Younger workers.

Of the twelve younger workers, six had been newly recruited to unionism on taking their job at the Polytechnic. Two were not and never had been members. One man refused categorically to respond to either of the union questions despite being exceptional in being a union member since leaving school. Perhaps the fact that when he was made redundant from both his earlier jobs he was 'sponsored' by his immediate superiors, rather than supported by his union, had contributed to his active disinterest.

The three people who had been union members in more than one job all selected their present job as the best for unionism and their reasons for their selections all illustrate the instability of their foothold in the labour market in their earlier employments:

1) The first case selected the present job because 'They now try and help us, before I just paid the money and thought "Who am I paying it to?"'. The reference to 'now' takes its meaning to the occasion in an earlier job when she had been promoted to a supervisor but was not been paid the increased rate. 'I kept asking the union to help me but they ignored me'.

2) This man's reason for selecting the present union was that 'The union actually fights for a small workforce here [i.e. his section]. In the other jobs I was never aware of the union'. He selected as his worst job for unionism his time as a driver/operator in a sewage contracting firm. This was a particularly unpleasant job but he dare not approach the union because of fear of losing the job. The present job compared 'very well' to that job because 'the union has a presence here'.

3) Here again the union has a 'presence' at the Polytechnic which contrasts with other experience. This woman selected the present job because 'The union is not prepared to let the employers change conditions of service...[Without the union] they would walk right over us'. This compared 'very well' with her worst union, in a section of local government where there was no branch organisation.

For a number of reasons the number of respondents replying to these questions was rather less than in the other aspects of jobs we have considered. Two people had never been union members and seventeen had only been members in one job and thus were unable to make comparisons. Another group of people refused to nominate one or more of the trade union questions. This chapter thus began with preliminary analyses of the reasons for non-membership and non-response before moving on to discussion of jobs nominated as best and worst for unionism. As in other aspects of jobs we did not find a clearcut pattern of association with industrial sectors, but once again the scheme of patterns of labour force participation, summarising the individual's work over time, proved useful in interpreting the jobs chosen as best and worst for unionism. The preoccupation of the career developers was the lack of fit between their occupational skills and the union which was recognised as representing them at the Polytechnic. For those who had entered the Polytechnic in response to an earlier employment crisis, the quality of shop floor representation was paramount, a criterion absent from those whose working life had been spent outside the context of large scale, highly unionised manufacturing establishments. In the case of women workers, concern was more particularised to specific experiences of trade unionism, often related to their status as women workers. The few younger workers in a position to compare different union practices all thought highly of the union at the Polytechnic as compared to union impotence in their earlier jobs. In each of these cases, best and worst jobs for unionism were not found systematically in particular industries or sectors but the main concerns of the individuals about trade unionism related to the overall pattern of their working lives.

Notes.

1. Similar pictures of the reasons for non-membership amongst sub-contractor cleaners and hotel and catering workers are reported in Coyle, A. 1986, Pp. 24 and Byrne, D. 1986, Chap. 6.
2. See, for example, the comments reported in Charles' study of women and shop-floor trade unionism. (Charles, N. 1986). A related point here is that the experience of unionism need not remain constant throughout a particular job. Thus some people identified the same work period as being both the best and worst for trade unionism. A particularly memorable example of this were several people who had worked at the 'Standard' plant (See chapter 5, footnote 9) for many years, who defined the unionism of the plant's heyday as their best and what they defined as the interminable meetings, sellouts and backstabbings of the dying days of the plant as their worst experience of unionism.
3. That is, union membership as a percentage of potential membership.
4. See Batstone, E. *et al* (1977), Brown, W. (1973), Friedman, A. (1977), Lyddon, D. (1983), Melman, S. (1957), Rayton, D. (1972), Tolliday, S. (1986), Turner, H.A. *et al* (1967), Zeitlin, J. (1980) & (1983).

5. The numbers here are simply for reference in the text.
6. One man did not know when he was least well served by a union.
7. See chapter 6 for details on the C.T.A..
8. This is reported in Chesterman, C.J. (1978).
9. For many years a major employer of female labour in Coventry, see O'Doherty, S.P. (1982) and Castle, J. (1986).

10 Best and worst jobs overall

We saw in chapter four that the people in the sample had considerable experience of work and employment in a wide variety of industrial contexts. The intention of the last set of questions was to ask respondents to nominate their best and worst jobs in overall terms. The questions asked to ascertain these were:

All things considered, which has been the best job you have ever had?

What was it about that job that, above all, you enjoyed?

So what you most enjoyed about that job was **KEYWORD**. *How does your present work compare to that job in this respect? Does it compare VERY BADLY/BADLY or is it NOT MUCH DIFFERENT?*

All things considered, which has been the worst job you have ever had?

What was it about that job that you especially disliked?

So what you most disliked about that job was **KEYWORD**. *How does your present work compare to that job in this respect? Does it compare VERY WELL/WELL or is it NOT MUCH DIFFERENT?*

The notion of 'all things considered' worked well and most people readily responded to these questions. The use of keywords in the comparison questions allowed the interviewer to focus the present job comparison to what was either most enjoyed or most disliked about the best/worst jobs overall. Only two people (both men, one an academic technician and one a protection officer) were unable to identify their best job, all things considered The distribution by industrial sector is shown in table 10.1.

Table 10.1
Best jobs overall
by industrial sector.

Sector	Number	Percent
Manufacturing	15	23.1
Distribution	6	9.2
Other services	8	12.3
Polytechnic	35	53.8
Military service	1	1.5
Total	65	100

What is immediately noticeable here is that just over half the sample identified a job at the Polytechnic as the best they had had. In most cases this was their current job, this applied in twenty-nine of the thirty-five cases in which a Polytechnic job was selected as best. Jobs in manufacturing come a long way behind with around a quarter of the best jobs. A further point of note is the scarcity of best jobs located in private services, only six jobs in distribution appearing here. When we turn to the industrial sector of worst jobs, a different picture emerges. These are displayed in table 10.2.

Table 10.2
Worst jobs overall
by industrial sector.

Sector	Number	Percent
Manufacturing	25	39.7
Construction	4	6.3
Distribution	12	19.0
Transport	1	1.6
Commerce	4	6.3
Other serv	6	9.5
Polytechnic	10	15.9
Unemployment	1	1.6
Total	63	100

Four people did not know (one male academic technician) or had never had (one female cleaner, one female technician, one male academic technician) a worst job 'all things considered'. The three people who had not had a worst job in this overall sense took this view because they had always enjoyed working and to identify a worst job, one suspected, was suggestive of complaint about their positive attitude to work.

Here the Polytechnic is selected in only ten cases, (only six of these chose their present jobs). The private service industries are more prominent with over a quarter of worst jobs. However, jobs in manufacturing and construction account for forty-six percent of the nominations. In terms of industrial sector it thus appears that best jobs were located in public services, especially the Polytechnic, and worst jobs in secondary industry and private services. In each case, however, a significant minority of a quarter to a third took the opposite view.

We can explore this a little further by relating these choices to the sex and current occupations of the sample members.

Table 10.3
Best jobs overall by industrial sector, sex and current occupation.

Sex & occup.	Manuf	Dist	Oth serv	Poly	Mil serv	Total
Women acad tech			2	6		8
Women cleaners	1	5	2	5		13
Caterers	6	1	2	3		12
All women	7	6	6	14		33
Men acad tech				9		9
Men maint tech				5		5
Protection	5		1	2	1	9
Men cleaners	2		1	3		6
Attendants	1			2		3
All men	8		2	21	1	32
Total	15	6	8	35	1	65

In terms of sex, fourteen of the thirty-three women in the sample thought a job at the Polytechnic their best, thirteen of them their present job. A rather higher proportion of men took this view, twenty-one of the thirty-two men who answered the question selected a Polytechnic job, though five of these chose a job prior to their present one. Turning to present occupation, the great majority of the technicians nominated a job at the Polytechnic, all five maintenance technicians took this view, as well as fifteen of the seventeen academic technicians who answered this question, though four of them chose an earlier job. Amongst cleaners, five of the thirteen women choose their present job, five others selecting a job in distribution. Again, half the male cleaners also chose a Polytechnic job as their best, two their present one. Two of the three attendants chose a Polytechnic job, one his present one. Much lower rates of selection of Polytechnic jobs occur amongst the caterers, three of them choosing their present job, and security staff, only two choosing their present job. Here the numbers selecting jobs in manufacturing are higher, six of the twelve caterers and five of the nine security men who made the choice. In terms of occupation, technicians select the Polytechnic as best overall, quite high proportions of cleaners and attendants make this choice with caterers and protection officers tending to select employment outside the Polytechnic. However, with the exception of the technicians there are significant exceptions to these trends.

The choice of worst jobs overall is displayed in table 10.4.

Table 10.4
Worst jobs overall by industrial sector, sex and current occupation.

Occupation	Man	Con	Dist	Tran	Comm	O serv	Poly	Unem	Total
Women acad tech	3		1			1	2		7
Women cleaners	4		3			3	2		12
Caterers	4		2		1	1	3	1	12
All women	11		6		1	5	7	1	31
Men acad tech	7					1			8
Maint tech	2	2	1						5
Protection	2		3	1	3		1		10
Men cleaners	2	1	1				2		6
Attendants	1	1	1						3
All men	14	4	6	1	3	1	3		32
Total	25	4	12	1	4	6	10	1	63

In terms of sex, it is noticeable that the choice of manufacturing jobs as worst of all is evenly spread between men and women and that seven women but only three men chose a Polytechnic job as their worst. Again in occupational terms, the bunching of technicians towards selecting manufacturing jobs is of note. However, aside from these the most prominent feature of table 10.4 is the lack of a distinctive pattern in these terms, in other words, within these occupational categories people selected a wide range of industrial sectors as the location of their worst jobs in overall terms.

The analysis of these choices shows a number of trends in terms of industrial sector and the sex/occupation of the respondents. However, in each case there are significant variations. We now turn to an analysis in terms of the work history of the individuals.

The career development work history pattern.

Of the eleven men in this group, ten selected their present job as their best all things considered. The exception selected a job when he worked as a research fellow in higher education, though he felt that this was 'not much different' to his present job. The strong trend toward selecting the current job is consistent with the pattern of work history which characterises these men, the development of a career in terms of the accumulation of skills in an occupational area which they have been able to transfer to the Polytechnic.

On the other hand, none of these men selected their present job as their worst in overall terms. One man refused to answer this question as he felt he had never had a worst job overall. The great majority chose a job very early in their working lives. In six cases this was the first job of all, in a further two the first civilian job after apprenticeship. In all these cases for which we have data[1] the present job compared 'very well' to these early jobs[2]. The two exceptions to this tendency to select very early jobs picked jobs in which they had experienced specific difficulties. One man was working as an industrial metallurgist and was put under enormous pressure to make key decisions concerning the tapping of furnaces without the necessary equipment or responsibility. Another was subject to technical change in the printing industry and felt that his so-called retraining in fact amounted to de-skilling.

The employment crisis work history pattern.

Perhaps rather surprisingly, two of these twelve men selected their present jobs as their best overall, one an attendant in a science laboratory who found the interest and responsibility of his present job much more to his liking than the thirty years he spent as a track assembly worker. The other has worked as a cleaner at the Polytechnic since 1971 when he was made redundant from a job in the aircraft industry. Quite why this man took and remained in his job as a cleaner is unknown but clearly from his nomination of the job as his best overall it is to his liking. One other man selected a cleaning job at the Polytechnic as his best overall, although not his present one. He was transferred from the job at his own request in what he now regards as a disastrous decision. So in the cases of three of these twelve, Polytechnic jobs, although of less skill and more poorly paid than their years in manufacturing, are regarded as their best all things considered. On the other hand nine of the twelve selected a job in manufacturing as their best overall, six of them the employment they held just before the period of employment crisis. The other three selected a job earlier than this, in each case a skilled job which they relinquished for higher paid work in Coventry's car factories.[3]

Turning to the worst job that these men had had, two identified their very first job as their worst, in each case these were low pay/low skill 'youth's' jobs prior to entering factory work. Three others identified a factory production job as their worst, one because the toolmaking shop were he worked was reorganised turning him into a mere machinist, one because of particularly poor working conditions in a foundry and the third selecting track assembly work as his worst. The biggest group here, however, selected a job in the process of transfer from these manufacturing jobs to their present occupations. Six men chose such a job, working for private security firms, in parcel delivery or as a car park attendant. Only one person selected his present job as his worst overall.

What is particularly noticeable here is the comparisons these men draw between their worst job and their present job, of the ten for whom we have data, eight thought the present job compared 'very well' and two 'well'.

The formerly mobile work history pattern.

One of the four men in this group was unable to nominate a best job overall, one selected an earlier period at the Polytechnic, though there was 'not much difference' between that job and his present one. The other two selected their present jobs. All four of them thought that their present jobs compared 'very well' to the worst jobs overall. We have noted earlier that as they moved about these men sometimes made 'mistakes' in the jobs they took, thus three of them stayed less than one year in their worst jobs, the other for one year only. They were not jobs held as youths nor in a transition process to their present jobs but occasions when a move 'went wrong', for example, one man trying his hand as a milkman, another choosing a job as a machinist when he emigrated to Australia.

The always services work history pattern.

One of these four men refused to answer either of these questions, for reasons which were not clear. The other three display little in the way of any pattern. One technician chose as his best job an earlier post similar to his present one and his worst a local government job which was the only time he moved out of the context of working in science laboratories. A cleaner nominated a period as a school caretaker as his best job and a spell as a commercial traveller as his worst. The third thought his present job as a security officer his best overall and working as a hotel chef his worst.

The women with a domestic interlude work history pattern.

In earlier chapters we have been able to relate these womens' selections of best/worst jobs to features of their work histories. In particular, the age when the job was held (closely related to pre/post domestic interlude), its full/part-time status, and industrial sector have been important. What is striking about tables 10.5 and 10.6 is the scatter of best/worst jobs 'all things considered' in terms of these variables.[4]

Table 10.5
Best jobs in overall terms.
(Post domestic women).

Number	Present job	Best job Occupation	Industry	F-T/P-T	Pre/Post dom break	Present job comparison
1	Caterer	Present		PT	Post	
2	Cleaner	Present		PT	Post	
3	Caterer	Present		FT	Post	
4	Cleaner	Present		PT	Post	
5	Cleaner	Present		PT	Post	
6	Caterer	Present		PT	Post	
7	Cleaner	Self-empl shopkeeper	Retail distribution	FT	Post	Badly
8	Caterer	Telephonist	Vehicle manufacture	FT	Pre	Not much diff
9	Caterer	Production clerk	Vehicle manufacture	FT	Pre	Badly
10	Cleaner	Vending machine operator	Vehicle components	FT	Post	Badly
11	Caterer	Machinist	Underwear manufacture	FT	Pre	Not much diff
12	Caterer	Component inspector	Electrical engineering	FT	Post	Not much diff
13	Cleaner	Barmaid	Social club	PT	Post	Not much diff
14	Cleaner	Shop assistant	Retail distribution	PT	Post	Not much diff
15	Cleaner	Present		PT	Post	
16	Acad tech	Present		FT	Post	
17	Cleaner	Secretary	Metal distribution	FT	Pre	Very badly
18	Caterer	Component inspector	Electrical engineering	PT	Post	Badly
19	Caterer	Buyer	Retail distribution	FT	Pre	Very badly
20	Cleaner	Registered childminder	At home	FT	Post	Not much diff.
21	Cleaner	Component inspector	Vehicle manufacture	FT	Post	Badly

Note, the numbers are simply for reference in the text.

Perusal of these tables shows that there is no simple pattern here. In terms of the full/part-time status of nominated jobs, nine women selected a part-time job as their best job, twelve a full-time job, but, on the other hand, eight people nominated a part-time job as their worst job, eleven a full-time job (one refused to chose a worst job overall). With reference to jobs located before or after the domestic interlude a similar picture emerges. Whilst only five women selected a job held prior to the break as their best job overall, as

Table 10.6
Worst jobs in overall terms.
(Post domestic women).

Number	Present job	Worst job Occupation	Industry	F-T/P-T	Pre/Post dom break	Present job comparison
1	Caterer	Machinist	Knitwear manufacture	FT	Pre	Very well
2	Cleaner	Kitchen assistant	Glass manufacture	PT	Post	Very well
3	Caterer	Shop assistant	Retail distribution	FT	Pre	Very well
4	Cleaner	Waitress	Restaurant	FT	Post	Very well
5	Cleaner	'Factory work'	Glass manufacture	FT	Pre	Very well
6	Caterer	Catering assistant	Motorway services	FT	Post	Very well
7	Cleaner	Meals supervisor	School meals	PT	Post	Very well
8	Caterer	Telephonist	Sub contracting engineers	FT	Pre	Very well
9	Caterer	Assembler	Electrical engineering	PT	Post	Very well
10	Cleaner	Cleaner	Retail distribution	FT	Post	Very well
11	Caterer	Nanny	Private household	FT	Pre	Very well
12	Caterer	Weaver	Ribbon manufacture	FT	Pre	Very well
13	Cleaner	Laundress	Hospital	PT	Post	Very well
14	Cleaner	Packer	Vehicle manufacture	FT	Pre	Very well
15	Cleaner	Component inspector	Electrical engineering	FT	Pre	Don't know
16	Acad tech	Research technician	Electrical engineering	FT	Pre	Don't know
17	Cleaner	Present		PT	Post	
18	Caterer	Present		PT	Post	
19	Caterer	Present		PT	Post	
20	Cleaner	Present		PT	Post	
21	Cleaner	Refused to say				

Note, the numbers are simply for reference in the text.

against sixteen choosing one after they returned to the labour market, eleven women selected a post-break job as their worst overall, nine nominating one held prior to the break. Again there is little in the sectoral distribution of jobs to differentiate the choices. Eight women selected a manufacturing job as their best, thirteen a service sector job but on the other hand nine people chose a manufacturing job as their worst, eleven others a service sector job. Eight people chose their present job as their best, but four selected it as their worst. It can also be seen that a number of occupations appear in both tables, that is, nominated as both best and worst jobs. This applies to catering assistants, machinists, component inspectors, shop assistants, telephonists and cleaners.

However, tables 10.5 and 10.6 do show a noticeable pattern in the comparisons the women make between present job and best/worst job. Taking the selection of best jobs first, eight women chose their present job as their best overall, six thought that there was 'not much difference' between the present job and the best job, five thought that the present job compared 'badly' and two 'very badly'. When we turn to worst jobs overall, a very different picture emerges. Only four of the women chose their present job as their worst job overall. One person refused to nominate a worst job and two did not know how the present job compared. But the remaining fourteen all thought that the present job compared 'very well' with the worst job they had held.

Clearly from both the point of view of best and worst overall the Polytechnic rates well. Two-fifths of the women selected the current job as the best overall, and the remainder did not draw strong comparisons between the present job and the selected best job. On the other hand, only a small number of women nominated the present job as the worst overall and the majority drew sharp comparisons between the current job and the worst. We can illuminate this contrasting pattern by using descriptive data supplied by sample members on the jobs chosen as best and worst.

As noted, fourteen of the twenty women who selected a worst job thought that the current job compared 'very well' to it. It was clear that such worst jobs did stand out in the women's work experience. The following case notes can illustrate this point. Number 3 selected a job as a shop assistant with a firm run by a father and son. The wives of the two men also came into the shop and she was bombarded with orders from all of them. Then, when two month's pregnant, she fell at work and took time off sick. Despite sending sick notes she received a letter informing her that she had been dismissed because of her 'condition'. Case number 6 identified her job as a catering worker in a motorway service station as her worst job. She worked a shift from 3 p.m. to 11 p.m. and had to travel some distance to and from work. Once there the pace of work was intense, 'we never stopped'. All in all she found the job exhausting and it was affecting her health. Number 15 related similar circumstances, working as a steakhouse waitress. She disliked the attitude of male customers to her, describing them as 'businessmen on expenses', but again the split shifts and changing shifts undermined her health. Number 10 is perhaps the most memorable, she was employed by a firm of cleaning sub-contractors, located at a large supermarket. The manager of the store told her to climb into a rubbish skip full of decomposing food to push it further to the back. She refused as she was employed to clean the store and went home. When she complained to her employer she was told that she had left the job voluntarily and a replacement had been advertised for. To cap it all she was then refused Unemployment Benefit for six weeks because of leaving a job for no good reason. This incident clearly stuck in this woman's memory. Whilst it is an extreme example of mistreatment the point is that in all these cases the worst job overall did stand out. By comparison with these experiences Polytechnic employment did compare 'very well'.

When we turn to case notes relating to jobs selected as best overall a degree of qualification has to be added to the favourable standing of the Polytechnic. It will be remembered that seven women thought that the present job compared 'badly' or 'very badly' to the job selected as best overall. These women felt trapped in their present jobs because of lack of opportunity to find alternative work. Most of them had had what they regarded as better jobs in the past but could not find comparable jobs at the present time. Case number 19, for example, had worked her way up to be a buyer in a department store, responsible for running a section of the store, including purchase of goods. She had left this position to have her family and now found herself, without qualifications, and unable to return to what had been a responsible job in which she had been successful.

Number 9 had worked as a clerk in the past and had recently taken a book-keeping course but was also unable to return to this occupation. Number 7 and her husband had bought a general store and run it successfully for a number of years until two large supermarkets had opened in the vicinity. The business became unviable and both her and her husband had had to take what they regarded as menial jobs. Here then were women who had had 'better' jobs in the past but were unable to continue in such positions.

Fourteen of the twenty-one women either selected the present job as best or saw 'not much difference' between the best job and current employment. However, six of these women were again constrained by circumstances. Number 16 felt overqualified for the job she was doing but had to continue because her husband was unemployed. Numbers 15 and 20 were hoping to take full-time employment when their children were older, in 15's case after returning to education. Numbers 6 and 11 had already done courses to try to obtain particular kinds of job, whilst number 2 was looking for full-time employment.

Thus, putting these two groups together, thirteen of the twenty-one women were either looking or hoping for something better. From the point of view of best jobs, the relatively favourable position of the Polytechnic arises from these women's lack of opportunity to pursue alternative employment. We should not forget, however, that from the point of view of worst jobs, the Polytechnic compares well. Thus the favourable position of the Polytechnic results from these womens' experience of poor jobs in the past and constraints on job possibilities at the present time.

Women with a non-domestic interlude work history pattern.

As in other respects there seems little to differentiate these two women from the rest. One, now working as a cleaner, selected a job as a hotel chambermaid, held in her teens, as her worst job and her present job compared 'very well' to it. Her best job was as a housekeeper, but this was 'not much different' to her present job. The other woman, a technician, selected as her best job an earlier local authority technician's post. Again there was 'not much difference' between this and her present job although this was selected as her worst overall in that she had been required to move to it because of personal circumstances. The general pattern of the Polytechnic rating well as a result of memorably poor jobs in the past and/or lack of opportunity in the present is thus reproduced here.

Younger workers.

Data on the jobs selected by the twelve younger workers in the sample are displayed in tables 10.7 and 10.8. It can be seen that in nominating their best jobs overall, six of the twelve selected their present positions. Of those that did not, three saw 'not much difference' between their present jobs and the chosen job and two others could not say, which in this context amounted to seeing little difference. Only one person felt that his present job compared 'very badly', he had been a career soldier with a skilled radio operating job in the army. On the other hand, none of the twelve selected their present job as their worst overall and eight said that their present job compared 'very well' to their worst job. So in these overall terms the familiar picture of the Polytechnic offering these younger workers a relatively favourable work environment, compared to other labour market experiences, is repeated.

Table 10.7
Best jobs in overall terms.
(Younger workers)

Number	Present job	Best job: Occupation	Best job: Industry	Present job comparison
1	Caterer	Caterer	Local government	Not much diff.
2	Security	Soldier	Military service	Very badly
3	Acad. tech.	Present		
4	Acad. tech.	Present		
5	Acad. tech.	Cartographer	Local government	Don't know
6	Acad. tech.	Technician	Polytechnic	Not much diff.
7	Acad. tech	Present		
8	Attendant	Painter	Polytechnic	Not much diff.
9	Cleaner	Present		
10	Caterer	Training scheme	Local government	Don't know
11	Acad. tech.	Present		
12	Maint. tech.	Present		

Note, the numbers are simply for reference in the text.

Table 10.8
Worst jobs in overall terms.
(Younger workers)

Number	Present job	Worst job Occupation	Industry	Present job comparison
1	Caterer	Unemployment		Very well
2	Security	Storeman	Wholesale distribution	Very well
3	Acad. tech.	Lab technician	Pharmaceuticals	Don't know
4	Acad. tech.	Lab technician	Hospital	Very well
5	Acad. tech.	Refused to say		
6	Acad. tech.	Temporary industrial artist	Electronic engineering	Very well
7	Acad. tech	Temporary shop assistant	Retail distribution	Very well
8	Attendant	Training course	Construction	Very well
9	Cleaner	Assembler	Electrical car components	Very well
10	Caterer	Caterer	Polytechnic	Not much diff.
11	Acad. tech.	Print room junior	Photographic laboratory	Well
12	Maint. tech.	Apprentice carpenter	Car components	Very well

Note, the numbers are simply for reference in the text.

Other.

The one case here selected her full-time job as her best job overall and her first job working as a secretary for a small family firm as her worst. Her part-time job at the Polytechnic was 'not much different' to the former but compared 'very well' to her worst job.

In terms of industrial sectors a high proportion of best jobs 'all things considered' were located at the Polytechnic with only a quarter within manufacturing. Conversely, the proportion of worst jobs overall found at the Polytechnic was relatively low with the secondary sector amounting to almost half and private services a further quarter. The significance of these sectoral patterns is amplified by analysis in terms of work history. Choices of best and worst jobs overall were consistent with the labour force participation of the career developers, those who had gone through an employment crisis, the formerly sectorally mobile and younger workers. Little pattern was found amongst those who had always worked in services. Amongst the older women workers a complex picture was revealed. On the one hand choices of best and worst jobs were scattered in terms of pre/post domestic interlude, full and part-time jobs and the industrial sector of jobs. On the other hand, comparisons with the present job did display a noticeable trend. This had two aspects, first, toward choice of current job as best or 'not much different' to the selected best job and, second, strong comparisons between present job and the job selected as worst. Descriptive casenote data have been deployed to amplify this trend in terms of the significance of memorably poor jobs in the past and lack of opportunity to move from the present job at present.

Notes.

1. In one case the item of information is missing.
2. In one case the actual response to the question of how the present and selected job compared was 'don't know' but the man made it clear that what he intended was that there was *no comparison* between the two jobs.
3. See the comments on this in chapter six.
4. Tables 10.5 and 10.6 should be read together, the identification numbers referring to the same individuals in each case.

11 Conclusion

The research problem and research site.

One of the most prominent aspects of the restructuring of employment in Britain in the last fifteen to twenty years has been the decline in the number of jobs in the secondary sector and the rise of employment in service industries. This report has been concerned with this structural shift but from a particular point of view; that of the individual worker who has lived through a quite dramatic period of economic change. The research project had two basic objectives. First to discover whether the *aggregate* shift in jobs from manufacturing and construction to services had involved the movement of *individuals* from one industrial sector to another? To the extent that individuals had moved, how had this occurred, what was the nature of 'sectoral transfer'? Second, insofar as people now working in services had previously been employed in the secondary sector, how does work compare in the two contexts?

To tackle these problems within the resources available to the researcher, a particular research site was required. The Coventry (Lanchester) Polytechnic was selected for three reasons. It was, first of all, located in a labour market which had undergone a rapid shift in employment in the recent past resulting from the plummeting of jobs in manufacturing and the steady increase in the service sector. Second, the Polytechnic had a degree of comparability to the large, unionised factory which had dominated the Coventry labour market but was located in an expanding service area. Finally, the Polytechnic employed a large number of ancillary workers and technicians for whom movement from the declining secondary sector was, in terms of skills and work experience, at least feasible. Had people working as cleaners, caterers, security staff, porters and attendants formerly worked in Coventry's vehicle, electrical and mechanical engineering plants? Had the Polytechnic recruited its maintenance and academic technicians from skilled workers in construction and engineering?

To answer these questions a research technique was adopted to document individuals' work history over the adult working life and to ask people to compare different jobs they had held. The structured questionnaire employed was designed to ascertain the individual's family history, work history and judgements about best and worst jobs. The family history was required to tackle the relationship between domestic commitments and paid employment, it covered the dates of marriage and various stages in the care of children. The work history established what work, in a broad sense, the individual had done since leaving the education system. For paid employment considerable detail on each job was collected to fill out the job comparisons made at the end of the interview. The latter were designed to get people to compare jobs they had held over the working life with reference to their currently held service sector employment. Respondents were asked to nominate their best and worst jobs and, insofar as the choice was not the present job, were further asked how that compared to the selected job.

Clearly, jobs can vary in different respects so that the questions asked here focused on key issues regarding secondary and service industry. All research into employment confirms that financial rewards are crucial. A second area of concern is 'skill' as the whole concept of skilled work originated and has its main manifestation in manual craft-work in the manufacture of things rather than the provision of a service. On the other hand, the typical place of manufacture is the factory, and by definition this involves the close supervision of work (Hobsbawn, E.J. 1969: Chap. 3). By contrast, service employment is generally little mechanised and is often not closely supervised. A third issue was thus job autonomy. The research site was highly unionised and was located in a city whose manufacturing workers had been in the forefront of the development of union workplace organisation in the post-war period. The individual's sense of trade union organisation was the fourth area of questioning here. Finally, respondents were asked to nominate their best and worst jobs 'all things considered' as a summary measure of their work experience.

At the time of the research field work, the Polytechnic employed 621 ancillary and technical workers, 345 women and 276 men. 310 people were employed part-time, 311 on a full-time basis. The sample of respondents was selected to represent the range of occupations mentioned above plus the gender and full/part-time composition of the workforce. In addition the data allows us to make some further observations on the kind of worker found in a workplace such as the Polytechnic. Workers tend to be relatively old, married and living in households with children. Their marital partners tend to be economically active. Sample members were recruited to the Polytechnic via informal means of job search and, once there, tend to stay for some time, although there is movement between jobs in the workplace.

Work history.

However, the main aim of the research was not a detailed study of the workplace but an investigation of the working lives of the respondents from school/college up to and including their present jobs. The average age at which workers had joined the Polytechnic was 36 years. Although there was some variation in this respect the fact is that the Polytechnic recruits people with considerable experience of work. The analysis of the aggregate work history, that of the sample as a whole, showed that this experience had been gained largely as employees. For the sample as a whole, 82% of adult life had been spent in employee status, the only other economic statuses of any overall significance being military service for men and unpaid domestic work for women. Seventeen of the men had spent some time in the forces, largely during wartime or National Service.

Twenty-one of the thirty-two women included in the analysis had been full-time 'housewives' but this time only amounted to a fifth of their total working lives.

Turning to the kind of employment the sample had held, a number of conclusions from the aggregate analysis suggest the transfer of individuals from the secondary to the service sector. First, 41% of their total working lives had been spent in manufacturing and construction and 24% at the Polytechnic. Second, a breakdown of when employment had been held in the different sectors showed a shift from secondary to service industry over the period from the 1940's to the 1980's. Third, fifty-six of the sixty-seven people had worked in the secondary sector at some point in their lives. But all these points, based upon the aggregate picture of the sample's work history, can only be suggestive in that the *sequence* of industrial participation for the *individual* cannot be ascertained. For example, it might be that people work in the secondary sector whilst young and then move to services for lighter work. Alternatively, people may move back and forth between the industrial sectors. Again, the aggregate analysis does not tell us when people move to the more minor (in aggregate terms) economic statuses, especially domestic work and unemployment. Thus the analysis has to shift from looking at the total working life of the sample or sub-samples to an examination of the course of individuals' working lives, how their participation in the labour force varies through adult life.

The working lives of sample members varied according to a number of factors, especially age, sex and the sequence of work experiences. We can use these factors to differentiate the patterns of labour market experience detected in the data. The significance of the worker's age can be introduced first. We noted earlier that the Polytechnic's workforce is relatively old and people generally join the workforce in their thirties or forties. Yet there was a significant minority of younger workers in the sample, that is, people in their twenties. There was only one teenager in the sample. Twelve of the respondents were aged thirty years or less and these had a pattern of work history distinct from that experienced by their older colleagues when they had first entered the labour market.

The younger workers had experienced considerable instability of employment since leaving the education system; nine of the twelve had been unemployed or participated in a government training scheme. By contrast, only five of the fifty-five older workers had been unemployed up to twenty-five years of age. Second, the younger workers had been gradually 'edging in' to the more secure employment that their present jobs represented. This took a variety of forms, an example is a young woman who, since leaving school had been unemployed five times, been on two 'training' schemes, spent a month employed in what she regarded as a 'sweatshop', and had been hired as a temporary clerk at the Polytechnic on two occasions. However, the latter represented a 'foot in the door' and as the second came to a close she was offered a permanent job on the basis of the reputation she had established. By comparison, the older workers had had very little difficulty finding employment and had moved from one job to another in their younger years. The third distinctive feature of the younger workers participation in the labour market was their lack of employment in manufacturing industry. As will be noted below, most of the older workers had spent considerable proportions of their early working lives in this sector. Of the twelve younger workers, five had no such experience and, in most cases, manufacturing employment amounted to a matter of months, often in 'special' circumstances such as part of a sandwich course whilst at college.

Turning now to the older workers in the sample, that is, those over thirty years of age. The next factor to introduce to discriminate between different work biographies is sex, in that men and women's working lives differed in three respects. A large proportion of the older women (twenty-one of the twenty-three) had interrupted their participation in paid employment for a 'domestic interlude', usually associated with the birth and care of young children. None of the men had done this. This break was an interlude, clearly by definition, all the women in the sample were now in paid employment, but overall they had spent only about a fifth of their working lives in full-time domestic work. The two exceptions to this pattern, older women with a continuous participation in paid employment can be kept separate for comparative purposes though we will see later that other 'domestic' constraints impinged on their lives.

When these women returned to employment after the domestic interlude it was to different kinds of jobs by comparison with their earlier work in terms of hours and industrial sector. The women had had sixty-nine periods of paid employment prior to the domestic interlude. Of these, sixty-six had been full-time jobs. In the post-interlude phase, the women had had eighty periods of paid employment, only twenty of these had been full-time, sixty part-time. None of the men had had part-time employment.

Turning, finally, to the location of jobs by industrial sector before and after the domestic interlude, the sixty-nine pre-interlude jobs had been split between thirty-six in the secondary sector and thirty-three in services. After the interlude only fourteen jobs had been in secondary industry; one was a period of continuous but casual employment in a range of sectors and the remaining sixty-five had been in the service sector. As we shall see many of the older men had much more continuous employment in manufacturing and construction, shifting later to the service sector.

The conclusion the evidence strongly suggests is the the older women's experience of secondary sector employment had been gained as younger workers. They 'transferred' to the service sector not directly as a result of the decline of manufacturing but on returning to paid employment after a period out of the labour market. At the time they still had considerable domestic obligations which led them to take part-time jobs which were located predominantly in the service sector.

When we turn to the older male workers, a further set of discriminations is necessary. Thirty-one of the thirty-four men in the sample fell into the over-thirty age-range. Of these, twenty-three had spent the great bulk of their working lives in the secondary sector before moving to service employment. However, the nature of this move to services followed two distinct patterns, the first we call the 'career development' group, the second the 'adjustment to employment crisis group'. Of the remaining eight older men, four had been sectorally mobile between industrial sectors over their working lives, the final four had spent all their civilian working life in service industry. We can take these in turn.

Eleven men fell into the career development group. They shared three characteristics. First, prior to the shift to service industry, almost all their working lives had been spent in secondary sector employment. In eight of the cases all civilian employment had been in either manufacturing or construction. So there was a clear move from one sector to the other. Second, this move can be characterised as a 'career development' because during working life these men had built up skills which they were able to take with them into service sector employment. Most were apprentice trained, some had obtained post-apprentice qualifications. On moving to service industry they had found jobs which required these skills. Thus, all had moved into a job which, in terms of its socio-economic ranking, maintained or enhanced their social position. Third, the process of transfer itself was not particularly associated with the 'shakeout' of labour in the early eighties and was not particularly traumatic for the individual. The year these individuals moved from the secondary sector varied from 1959 to 1982. Only one man spent time unemployed in the process, and that lasted only one month. It is of note, however, that in seven of the eleven cases the reason for the move was anticipation of redundancy, actual redundancy or deskilling. In other words, these men were 'pushed' into sectoral transfer by wider economic forces. The crucial point is that they were able to respond to this push in a way beneficial to themselves by deploying their skills in a different working environment.

This is the main point of contrast with the next group of twelve men, those who had had to adjust to an employment crisis. Like the career developers these men had spent most of their pre-service working lives in the secondary sector. Most of them had finished this phase of life with a long spell in one of Coventry's car plants, the average time with the last secondary sector employer was twenty years! All of them had been made redundant from these jobs, at which point they had begun a traumatic period of life which had eventually led to their present employment. The trauma had involved long spells of unemployment, temporary jobs, often with what were regarded as 'cowboy' firms, and 'retraining' in skills of peripheral relevance. At the end of this process they had obtained employment at the Polytechnic in jobs inferior to their earlier work but far more satisfactory than the intervening period. An example is a man who was made redundant from the British Leyland Canley plant in 1979 after fourteen years as an assembly track fitter. He was then unemployed for nine months, had a temporary warehouse job for five months, was unemployed again for seven months, went on a government training scheme in welding for six months, returned to being unemployed for two years nine months before securing a job as an attendant at the Polytechnic in the autumn of 1984.

The common feature of the men so far described is that they had begun their working lives in the secondary sector and spent considerable time there before transferring to services. The next group of four men had been sectorally mobile over their working lives moving from job to job across industrial sectors. The point to note here is that this mobility came to a halt in the mid-seventies. A number of factors are possible explanations for this change in work history pattern but the most plausible in the light of the evidence is that the tightening labour market of the later seventies precluded the job mobility these men had pursued for periods of fifteen to twenty years before remaining at the Polytechnic for an average of ten years.

This leaves four men who had never worked in the secondary sector, spending all their civilian employment in the service sector.[1] The point to note here is the small number of cases here. For most of the older workers some sense of 'transfer to services' applied. However, the nature of this process differed. In the case of older women, the move to service industry was connected to life-cycle stage rather than a direct result of the decline of manufacturing. It is, however, the case that the availability of part-time employment is intimately connected to the growth of the service sector so there is an indirect link here. For some of the older men the move to service employment represents a career development, often precipitated by redundancy from the secondary sector, though this was not limited to the steep decline in manufacturing of the last ten years or so. For other men their displacement from secondary sector jobs was traumatic, a result of recession and led to sectoral transfer as a way out of a period of employment crisis even though their present jobs were in many ways[2] inferior to those they had formerly held in secondary industry. The notion of 'sectoral transfer' did not apply in the case of younger workers in the sample as their experience of secondary sector employment was minimal.

Best and worst jobs.

These then are the main patterns in the sample's work history with reference to the question of the move from secondary to service industry. The second concern of the research project was to investigate people's views on their present jobs by comparison with earlier work. Chapters six to ten provide detailed analyses of the choices of best and worst jobs in five aspects: money, skill, job autonomy, trade unionism and 'all things considered'.

They also consider how present jobs fared in these selections and compared to nominated jobs. In each case the analysis centered on the interpretation of these views in terms of the patterns of labour market experience above. Rather than simply summarising this in terms of the five job aspects we will here assemble the main conclusions in relation to the different work biographies just described. From this we can gain a sense of what the various trajectories of work over time mean to the individual in terms of the five aspects of work investigated.

The career development work history pattern.

From the financial aspect the career developers generally selected their present job as their best and a job held early in life as their worst. There were some noticeable exceptions to this. In some cases the jobs these men had had to leave in manufacturing were their most financially rewarding. In two cases the move to services had amounted to a trade-off between money and maintaining skill as the present job was regarded as the worst for money but was taken to maintain the skill-content of the person's work.

In skill terms the present job was rated highly in either being chosen as the most skilled or being 'not much different' to the most skilled. Two exceptions here are men who selected periods of self-employment as their most skilled, to which their present jobs compared 'badly'. This seems to be a significant variation from the overall pattern of accumulating skills over a work history of career development. However, none of the men chose their current jobs as their worst in skill terms. Some refused to admit ever having such a job, others located their worst job early in working life. All who selected a worst skilled job thought their present job compared 'very well' to it.

In job autonomy terms most of the men in this group selected their present jobs as the most autonomous they had held and a secondary sector job as their least autonomous, drawing strong comparisons between the latter and their present jobs. However, in most cases the least autonomous job was their apprenticeship and it was as trainees rather than as secondary sector workers that the contrast was drawn.

The notion of career development thus generally describes these men's views as regards money, skill and job autonomy. It is not very applicable in the case of their experience of trade union representation. Those answering these questions did not feel well represented by their present union. This was because their union identification was with earlier craft unions representing their occupations. Their present 'industrial' union was perceived as detached from their concerns as technical and craft workers.

Looking at jobs in overall terms ten of the eleven chose their current job as the best they had had and the one exception's present job was 'not much different' to his best. Most selected early jobs as their worst 'all things considered'. Thus in four of the five job aspects the work history pattern is sustained by the men's views about the jobs they have held. From both points of view their working lives form a trajectory of improvement in which the main points of comparison are between present jobs and ones held early in life whilst skills were being learnt, skills which have served them well. The exception to this is trade unionism in which their current membership of NALGO is not perceived as relevant to men brought up in craft union traditions of occupational representation.

The employment crisis work history pattern.

Only one of this group identified his present job as his best in money terms, the rest all selecting earlier manufacturing jobs, the majority the long standing job they had held prior to redundancy. Furthermore, the present job compared unfavourably financially to these best jobs. In four of the twelve cases the present job was selected as the worst for money, three others selecting a job held during the transfer process with the remaining five largely selecting jobs held in their youth. So from the point of view of both best and worst choices the move to service sector employment has not been financially beneficial for these workers.

A similar picture emerges when we turn to these men's selection of most and least skilled jobs. None of the group chose his present job as his most skilled, rather earlier manufacturing jobs were nominated and strong comparisons drawn between these jobs and present employment. Five of the men thought their present jobs their least skilled, six others choosing a job in the process of transfer from manufacturing to their current positions.

In terms of job autonomy a rather complicated pattern was revealed. Whilst four men thought their present jobs gave them most autonomy, a further six selected a manufacturing job. However, in these cases the present job was 'not much different' to the chosen job. On the other hand, the present job compared very favourably to jobs selected as least autonomous. So in this respect, although no simple contrast emerged between the constraints of the factory and the autonomy of the Polytechnic, the latter did compare rather more favourably than in money and skill terms.

In terms of trade union representation there was again no simple contrast between present service sector work and earlier manufacturing experience. Whilst seven men chose a manufacturing job as their best for unionism, three chose the present job. Again only three felt that they were least well served by their present trade union, the majority selecting jobs in manufacturing. However, what does come out strongly from the responses to these questions is the men's concern with the quality of workplace representation as their criterion of being well or poorly served by a trade union. This was not the preserve of either the factory or the Polytechnic exclusively but might be argued to be the product of these men's long experience of strong shop-floor organisation in Coventry's manufacturing plants.

Turning to best and worst jobs 'all things considered', the majority selected an earlier manufacturing job as their best. However, three men thought their present job to be the best they had held and only one chose his job as his worst overall. The most prominent choice of worst job was one held in the transition from manufacturing to the present position. Thus, in these overall terms, the Polytechnic comes out rather well in view of the decline in these men's earnings and skills used and the rather mixed picture as regards job autonomy and trade unionism. The reason for this is that although the present job represents a deterioration as compared to best jobs held, it is by no means the worst that these men have experienced. This is indicated by looking at the comparisons they draw with worst jobs. In each of the five job aspects covered there is a tendency for the Polytechnic to compare 'very well' to worst jobs. Many of the latter were held during the traumatic period of transition between redundancy from manufacturing and securing the present job. Whilst the Polytechnic compares unfavourably with these men's best jobs in manufacturing it compares relatively favourably to other available alternatives, alternatives which many experienced en route to obtaining their current employment.

The formerly mobile work history pattern.

In money terms two of these four men selected earlier, manufacturing jobs as their best for money, two their present jobs despite apparently holding better paid jobs in the past which do not appear to have remained firmly in their memories. What have are their worst jobs for money, times when they made 'mistakes' in their many movements between jobs and entered positions which lasted a short time and to which the present job compared very favourably.

Turning to the skill aspect, none of the men chose their present jobs as the most or least skillful jobs. Their most skilled jobs were when they were employing the training they had received, but the present job was not perceived as being much different to this. On the other hand, their least skillful jobs were when they were not working in their 'trade' and work at the Polytechnic compared very well to this.

This pattern is repeated for job autonomy, little difference was perceived between the best job in this respect and the present job, whilst strong contrasts were drawn between the least autonomous job and the present one. Here least autonomous jobs were found in the secondary sector, most autonomous in the service sector. Thus amongst men who moved from job to job the contrast of the factory and services seemed apparent.

The data on trade union representation was very similar to that for those adjusting to employment crisis, best and worst jobs distributed across different industrial contexts with a concern for workplace representation coming through in the open ended questions used to probe the responses.

Looking at jobs 'all things considered', all of these men chose a worst job to which their present job compared very favourably. These were jobs held for brief periods of time in a varied career in which many things had been tried and mistakes occasionally made. By comparison with these the Polytechnic stood out well. On the other hand, although the Polytechnic might not be the best job these men had had in the various aspects asked about, it did not compare particularly unfavourably. For example, in skill terms, most of them had been trained in a particular occupational area but had not consistently pursued a career in that particular occupation.

Prior to settling at the Polytechnic these men had moved between occupations and industries. Whilst this gave them a wide variety of experience they lacked the main point of reference of the first two groups we have discussed, having neither the consistent employment in a particular skilled occupation of the career developers nor the stable, long term position with one employer of the adjusters to crisis. So in response to the best job questions we find a mix of jobs and insofar as the present job is not selected it is not perceived to be much different. On the other hand, worst jobs were very identifiable and the present job compared very favourably. The curtailment of job mobility had thus been reasonably satisfactory from these mens' points of view. The Polytechnic represented markedly superior conditions to the worst jobs they had held and was not much different to the best jobs in their experience.

The always services work history pattern.

In terms of the experience of these men the Polytechnic paid relatively good wages, all four choosing their present job as financially their best. Their worst jobs for money had been in their early years in the civilian labour market, to which their present jobs compared 'very well'. Similarly, in terms of job autonomy, the Polytechnic fared well by comparison with other service sector jobs. Most thought their present job their most autonomous, none felt it was their least autonomous and all thought their current job compared 'very well' to their worst job in this respect.

In skill terms the Polytechnic came out rather less well, none of the four thinking their present job their best in this respect and one considering it his least skilfull. On the other hand two thought the current job to compare 'very well' to their least skilfull work. As far as trade unionism was concerned these men were rather negative in general, but tended to view matters in rather more 'abstract' terms than the earlier groups we have mentioned. For example, two had resigned from their union on matters of principle.

'All things considered' a range of jobs were selected in both best and worst respects. Although the Polytechnic scored well financially and in terms of job autonomy this did not lead it to be selected as best overall. Neither did its mediocre rating on skill and trade unionism mean that it was chosen as worst. Thus in these overall terms little pattern emerged.

Women with a domestic interlude work history pattern.

Most of the twenty-one women here were part-time workers. It was thus something of a surprise to find that twelve of them regarded their current job as their best for money and that most of those who chose another job saw 'not much difference' between it and the present position. Conversely, none of the women selected the present job as the worst for money and many felt their Polytechnic job to compare 'very well' to their worst. The points of reference here are of interest. Those who did not chose the Polytechnic as their financially best job tended to chose times when, as adults, they had held full-time jobs in manufacturing. Insofar as an individual had not had such a position then the Polytechnic was nominated. Worst jobs for money were held either in the woman's youth or were part-time jobs in other service industries. The Polytechnic is thus regarded favourably in terms of these women's experience of, on the one hand, poorly paid work in the youth labour market and part-time service sector employment and, on the other hand, full-time adult work in manufacturing.

A similarly favourable picture emerged from the data on job autonomy. Thirteen of the twenty-one thought their present job their most autonomous and those that selected another job did not draw strong comparisons between it and the present work. None of the group considered their present job to be their least autonomous and many felt that it compared 'very well' to their worst jobs in this respect.

In skill terms, however, things looked very different. Only one woman thought her present job her most skilfull, the rest generally choosing full-time jobs held in their twenties, spread across industrial sectors. The present job compared unfavourably to these, in eighteen cases it was thought to compare either 'badly' or 'very badly'. Again, eleven of the women considered their current job to be their least skilfull, most of the remainder choosing other part-time service sector jobs held after their domestic interlude.

As far as trade unionism is concerned, one third of these women had been recruited to unionism for the first time at the Polytechnic. For those in a position to compare different experiences of trade unionism a high proportion, eight of the fourteen, related instances of poor treatment by unions or their officials, often connected to their status as *women* workers. Such instances, however, seemed to be spread across the life-cycle and industrial sectors. The Polytechnic seemed no better and no worse than other union involvements these women had had.

Thus the present work situation of these women compared favourably to others in financial and job autonomy terms, very poorly in skill terms and was much of a muchness as far as trade unionism was concerned. Perhaps this is why their best and worst jobs 'all things considered' were scattered across the age range, the family life-cycle, full and part-time status and industrial sector. However, the women's comparisons between their present jobs and their best/worst jobs overall were instructive. Only four women thought the present job to be the worst they had had overall. Most of the remainder, fourteen, said their present jobs compared 'very well' to their worst and we were able to illustrate the meaning of this statistic by the instances of mistreatment that many of the sample related to us. The Polytechnic fares very well by such comparisons. Again, eight of the women thought the Polytechnic to be their best job overall and a further six felt there was 'not much difference' between their best job and the current one. Yet at the same time it was clear that many of these women felt constrained doing their current work. Thirteen of them were either looking or hoping for something better when either their circumstances changed or employment opportunities improved. Whilst from the point of view of comparisons with worst jobs the Polytechnic is a good employer, from the point of view of best jobs it fares well because of the lack of preferred alternatives for these workers.

Women without a domestic interlude work history pattern.

Two women were distinguished because they had not taken a 'domestic interlude' from employment, having been in paid jobs all their working lives. Yet in all respects the pattern of their responses did not differ from the post-domestic women just described. For example, their present jobs were rated poorly in skill terms and their choices of best/worst jobs overall displayed the factors of memorably poor jobs and lack of opportunity outlined above. In other words neither had been able to accumulate deployable skills despite the lack of a career break. The reason for this was quite clear. Other 'domestic' circumstances had intervened in a way parallel to withdrawing from the labour market. One woman had left her nursing job to care for a sick sister, the other had had to leave her best job overall because of a marriage rule. This only serves to show that women's domestic obligations inhibit their role in the labour market in a multiplicity of ways not confined to the bearing and care of children.

Younger workers.

The labour market experience of the younger workers in the sample was marked by instability and a gradual establishment of the more secure positions their present jobs represented. From this one would expect the Polytechnic to be regarded favourably in each of the respects considered here. This is indeed the case. In monetary terms eleven of the twelve chose the present job as the best and none as the worst. In skill terms those employed as technicians thought their present job their best, the rest saw 'not much difference' between current job and the most skillful work they had done. Only one regarded the present job as the least skillful, the rest drew strong comparisons between present and least skilled job. Ten chose their Polytechnic job as their most autonomous, only one person considering the present job as the least autonomous, the majority feeling their jobs compared 'very well' to those nominated.

Two of the younger workers had never been union members, six had been recruited to unionism at the Polytechnic. Of the remaining four, one was generally antagonistic to unionism but the other three all thought their present union their best, comparing its 'presence' at the Polytechnic with earlier senses of union impotence. In overall terms six felt their present jobs to be their best and most of the rest saw 'not much difference' between their best jobs and their current one. None felt their present job to be their worst overall, most again seeing their present jobs as comparing 'very well' to their worst. Generally then, in terms of the experience of these workers the Polytechnic is understandably rated highly.

Work history and attitudes to work in the service sector.

The pattern of response to the various questions on best and worst jobs thus makes sense in terms of these workers' patterns of participation in the labour market over their working lives. However it is crucial at this point to bear in mind that both the work history patterns and the estimations of best/worst jobs have been derived from data drawn from a particular workplace with specific characteristics. The Polytechnic was chosen as a large, public sector establishment with full trade union representation and a range of manual occupations. As has been mentioned earlier the workforce tends to be relatively old and relatively long serving with the employer. Sample members lived in stable familial situations with economically active partners. They had obtained their jobs through informal means suggesting a network of contacts with those already in this kind of employment.

For all these reasons the results of this project cannot be regarded as typical of work in the service sector as a whole or even to workplaces or occupations which are similar in some respects. Hotels are sometimes large establishments employing high proportions of part-time, female labour. But as Byrne's (1986) study shows trade union representation in the hotel industry is limited, workers tend to be relatively young, often from ethnic minorities and employment is extremely unstable. Again, private sector cleaning companies employ part-time female labour and according to our respondents 'cleaning is cleaning'. But Coyle's (1986) report on sub-contracting cleaning work suggests many differences in the workforce and conditions of work as compared to that reported here. In other words we must insist that the research results reported here are particular to the labour market, workplace and workforce studied.

What this case study has sought to establish is two things:

a) First of all the viability and importance of an individual work history approach to economic change. All the workers in this study have experienced the dramatic changes in the labour market over the last fifteen years or so. Yet its meaning to them varies according to their pattern of participation in the labour market over their working lives. We have sought to empirically analyse such patterns in chapters four and five of this report.

b) Second, we have sought, in chapters six to ten, to show the importance of setting individuals' views of their present position in the context of their experience over time. A couple of examples can illustrate this. In our 'adjusters to employment crisis' group were men who had spent many years of their lives in some of the highest paying, well organised manufacturing plants in the motor industry. By comparison their present wages were not high. Yet their views must be placed in the context of what we have here called the trauma of the transfer process. The relatively favourable position of the Polytechnic is understandable as compared to the experience of unemployment, temporary jobs and retraining which marked the transition. Again, by most standards the part-time women cleaners and caterers at the Polytechnic are not highly paid. Yet we found that they felt themselves to be relatively well off *in comparison with other employment they had had.*

It is hoped that the results of this particular case study will thus generate further interest and understanding of how individuals experience economic change, especially in the growing and diverse service sector industries and especially in the context of working lives seen over time.

Notes.

1. There is also one other case, a woman who had a part-time job at the Polytechnic at the weekend which was supplementary to her main job. She was a special case not only in this respect but also in that her main job was in manufacturing industry so that she had not transferred to services.
2. We shall turn to the respondents' views on present and past jobs in a moment.

Bibliography

Allen, S. (1982) 'Women in Local Labour Markets' in J. Laite (ed) **Bibliographies on Local Labour Markets and the Informal Economy** London, S.S.R.C.

Barrere-Maurisson *et al* (1985) 'The Course of Women's Careers and Family Life' in B. Roberts *et al* **New Approaches to Economic Life** Manchester, Manchester University Press, Pp. 431-458.

Batstone, E. *et al* (1977) **Shop Stewards in Action, the Organisation of Workplace Conflict and Accomodation,** Oxford, Blackwell.

Blackburn, R.M. & Mann, M. (1979) **The Working Class in the Labour Market** London, MacMillan.

Brown, R.K. (1984) 'Working on Work' **Sociology** 18, 3, August, Pp. 311-323.

Brown, W. (1971) 'Piecework Wage Determination in Coventry' **Scottish Journal of Political Economy** 18, February, Pp. 1-30.

Brown, W. (1973) **Piecework Bargaining.** London, Heinemann.

Byrne, D. (1986) **Waiting for Change? Working in Hotels and Catering.** London, Low Pay Unit and GMBATU.

Castle, J. (1986) 'Factory Work for Women: Courtaulds and G.E.C. Between the Wars.' in B. Lancaster and T. Mason **Life and Labour in a twentieth Century City** Coventry, Cryfield Press, Pp. 133-171.

Charles, N. (1986) 'Women and Trade Unions.' in Feminist Review **Waged Work: A Reader.** London, Virago.

Chesterman, C.J. (1978) 'Women in Part-Time Employment: An Investigation of the Growth of Part-Time Employment in Britain since World War II with Particular

Reference to Patterns of Employment in Coventry.' M.A. Dissertation, University of Warwick Library.

City Treasurer's Economic Unit (1986) 'Changing Working Patterns and the Role of Manufacturing' **City of Coventry Economic Monitor** No. 2, June, Pp. 43-47.

Cockburn, C. (1986) 'The Material of Male Power' in Feminist Review **Waged Work: A Reader** London, Virago, Pp. 93-113.

Coventry City Council (1930) **The City of Coventry as an Industrial, Commercial and Residential Centre.** Cheltenham, E.J. Burrow and Co.

Coyle, A. (1986) **Dirty Business: Women's Work and Trade Union Organisation in Contract Cleaning.** Birmingham, West Midlands Low Pay Unit.

Finn, D. (1984) 'Leaving School and Growing Up: Work Experience in the Juvenile Labour Market' in I. Bates et. al. **Schooling for the Dole** Basingstoke, MacMillan.

Hakim, C. (1982) 'The Social Consequences of High Unemployment' **Journal of Social Policy** 2, 4, Pp. 433-467.

Gershuny, J.I. and Miles, I.D. (1985) 'Towards a New Social Economics' in B. Roberts *et al* **New Approaches to Economic Life** Manchester, Manchester University Press, Pp. 24-47.

Goldthorpe, J. *et al* (1968) **The Affluent Worker: Industrial Attitudes and Behaviour.** Cambridge, Cambridge University Press.

Handy, C. (1984) **The Future of Work** Oxford, Basil Blackwell.

Healey, M. and Clark, D. (1983) 'Industrial Change in Coventry: 1974-82' **City of Coventry Economic Monitor** No. 4, December, Pp. 51-66.

Healey, M. and Clark, D. (1984) 'The Coventry Region Industrial Establishment Databank I: A Guide to Sources, Methods and Definitions used for the City of Coventry' Coventry (Lanchester) Polytechnic, Department of Geography, Industrial Location Working Paper No. 2, December.

Hinton, J. (1973) **The First Shop Stewards' Movement.** London, George Allen and Unwin.

Hobsbawn, E.J. (1969) **Industry and Empire.** Harmondsworth, Pelican.

Jones, B. (1982) **Sleepers Awake! Technology and the Future of Work.** Brighton, Wheatsheaf.

Knowles, K.G.J.C. and Robinson, D. (1969) 'Wage Movements in Coventry' **Bulletin of the Oxford University Institute of Economics and Statistics** 31, 1.

Lyddon, D. (1983) 'Workplace Organisation in the British Car Industry.' **History Workshop** 15, Spring, Pp. 131-140.

Martin, J. and Roberts, C. (1984) **Women and Employment: A Lifetime Perspective** London, H.M.S.O.

Massey, D. (1984) **Spatial Divisions of Labour** London, MacMillan.

McG. Davies, J, (1986) 'A Twentieth Century Paternalist: Alfred Herbert and the Skilled Coventry Workman'. in B. Lancaster and T. Mason **Life and Labour in a Twentieth-Century City: the Experience of Coventry** Coventry, Cryfield Press, Pp. 98-132.

McKee, L. & Bell, C. (1985) 'Marital and Family Relations in Times of Male Unemployment' in B. Roberts et. al. **New Approaches to Economic Life** Manchester, Manchester University Press, Pp. 387-399.

NALGO News (1986) 'Poverty in Universities' No. 247, 26th September, Pp.4-5.

O'Doherty, S.P. (1982) 'A Study of Union Organisation and Effectiveness within the G.E.C. Telecommunications Plant, Coventry.' M.A. dissertation, University of Warwick library.

O.P.C.S. (1980) **Classification of Occupations** London, H.M.S.O.

Pahl, R.E. (1980) 'Employment, Work and the Domestic Division of Labour' **International Journal of Urban and Regional Research** 4, 1, Pp.1-20

Pahl, R.E. (1984) **Divisions of Labour** Oxford, Basil Blackwell.

Phillips, A. and Taylor, B. (1986) 'Sex and Skill' in Feminist Review **Waged Work: A Reader** London, Virago, Pp. 54-66.

Price, R. and Bain, G.S. (1983) 'Union Growth in Britain: Retrospect and Prospect.' **British Journal of Industrial Relations** 21, 1.

Procter I. (1984a) 'Employment Patterns in Coventry, 1911-71' Unpublished working paper, University of Warwick, Dept. of Sociology, March.

Procter, I. (1984b) 'Coventry as Compared to Other British Cities' Unpublished working paper, University of Warwick, Dept. of Sociology, December.

Procter, I. (1984c) 'A Review of the 1981 Census Economic Activity Statistics for Coventry'. Unpublished working paper, University of Warwick, Dept. of Sociology, November.

Richardson, K. (1972) **Twentieth-Century Coventry** London, MacMillan.

Robertson, J.A.S. *et al* **Structure and Employment Prospects of the Service Industries** London, Department of Employment, Research Paper No. 30.

Rosser, M. and Mallier, T. (1981) 'The Economic Base of Coventry' University of Leicester, Public Sector Economic Research Centre, Discussion Paper L81/05.

South, N. (1982) 'The Informal Economy and Local Labour Markets' in J. Laite (ed) **Bibliographies on Local Labour Markets and the Informal Economy** London, S.S.R.C.

Thompson, P. (1978) **The Voice of the Past** London, Oxford University Press.

Thompson, P. (1983) **The Nature of Work** London, MacMillan.

Thoms, D.W. and Donnelly, T. (1986) 'Coventry's Industrial Economy 1880-1980' in B. Lancaster and T. Mason **Life and Labour in a Twentieth-Century City: the Experience of Coventry** Coventry, Cryfield Press, Pp. 11-56.

Tolliday, S. (1986) 'High Tide and After: Coventry's Engineering Workers and Shopfloor Bargaining, 1945-80.' in B. Lancaster and T. Mason **Life and Labour in a Twentieth-Century City: the Experience of Coventry** Coventry, Cryfield Press, Pp. 204-243.

Turner, H.A. *et al* (1967) **Labour Relations in the Motor Industry.** London, George Allen and Unwin.

Whitston, K. (1979) 'The National Union of Vehicle Builders Between the Wars: Craft Unionism and Mass Production.' M.A. dissertation, University of Warwick Library.

Zeitlin, J. (1980) 'The Emergence of Shop Steward Organisation and Job Control in the British Car Industry.' **History Workshop** 10, Autumn, Pp. 119-137.

Zeitlin, J. (1983) 'Workplace Militancy: A Rejoinder.' **History Workshop** 16, Autumn, Pp. 131-136.